Listening to Sea Lions

Listening to Sea Lions

Currents of Change from Galapagos to Patagonia

Sarah Keene Meltzoff

ALTAMIRA
PRESS
A division of
ROWMAN & LITTLEFIELD PUBLISHERS, INC.
Lanham • New York • Toronto • Plymouth, UK

Published by AltaMira Press
A division of The Rowman & Littlefield Publishing Group, Inc.
A wholly owned subsidiary of The Rowman & Littlefield Publishing Group, Inc.
4501 Forbes Boulevard, Suite 200, Lanham, Maryland 20706
www.rowman.com

10 Thornbury Road, Plymouth PL6 7PP, United Kingdom

British Library Cataloguing in Publication Information Available

Library of Congress Cataloging-in-Publication Data
Meltzoff, Sarah K.
Listening to sea lions : currents of change from Galapagos to Patagonia / Sarah Keen Meltzoff.
p. cm.
Includes bibliographical references and index.
ISBN 978-0-7591-2235-2 (cloth : alk. paper) — ISBN 978-0-7591-2237-6 (ebook)
Human ecology—Pacific Coast (South America) 2. Marine ecology—Pacific Coast (South America) 3. Coastal ecology—Pacific Coast (South America) 4. Climatic changes—Pacific Coast (South America) 5. El Ni?o Current 6. Pacific Coast (South America) —Environmental conditions. I. Title.
GF531.M45 2012
304.2098—dc23
2012032336

The paper used in this publication meets the minimum requirements of American National Standard for Information Sciences Permanence of Paper for Printed Library Materials, ANSI/NISO Z39.48-1992.

Printed in the United States of America

To dearest Annie Laurie

Contents

Preface

"PUTTING THE CART BEFORE THE HORSE": DOING FIELDWORK IN THE SOLOMONS

Perhaps because I'd spent so much time in the Mayan highlands of Mexico and Guatemala, I won a T. J. Watson Fellowship during my final year at Connecticut College. I ended up in Solomon Islands villages fringed by coral reefs where ritual hunters captured tuna and porpoise.

The Solomons, however, hadn't been my original Watson proposal. At the time of my Watson application, I was taking classes in celestial navigation and piloting at the U.S. Coast Guard Academy. I imagined sailing solo around the globe and dreamed I could do it. So I'd proposed to the Watson people sailing solo down the west coast of South America, visiting coastal villages to observe children's play, transposing my Mayan village experience to my love of the ocean. I won the Watson. Near the end of the academic year, my Coast Guard professor presented me with the region's HO book, containing exact maps and charts of South America's Pacific coastline.

Examining the charts, I realized that the stretch of coast I'd fantasized as dotted in quaint villages was desolate desert. My heart sank in the reality of cold Humboldt currents, upwellings, rough seas, and dangerously rocky coasts. I asked the Watson people if they minded exactly where I did field research. They reassured me, "Not as long as you work overseas independently."

I landed in the Solomon Islands with the Watson's blessing at the beginning of 1973, as far away as geographically possible: halfway around the globe, twelve hours time difference, south of the equator. The Solomons was still a British colony; nobody at the UN could send me a brochure of what to expect.

Those Solomons of 1973 have entered my dream space. I lived in leaf houses in Small Malaita with no roads or electricity. I can still smell the fringe reef shore, feel myself paddling canoes to travel, and recall the excitement stirred by the outside world in the form of an occasional cargo boat. Village life was surrounded by spirits such as the Akaloni Matawa—the Deep Sea Spirit—with tuna fish as hands, feet, and head. My favorite is still Loloangili, a Makira chief's daughter captured by an army of ancestral sharks whose spirit escapes into her pet flying fox—a giant fruit bat—when a regular shark eats her. Initially, I'd been greeted with salutes, "Hello Joe, America come now! New York come now!"

Village life in the Solomons was wrapped tightly in ocean and rain forest spirits, led by the missionaries' Jesus-as-top-god. I joined an island world populated by spirits and ancestors who remain active in affairs as the "living dead," who brought success or failure, health or disaster. Living in isolation from the outside, sensing intense cultural differences, building close friendships as I was becoming part of another world—I was undergoing a classic anthropological initiation.

Adults described Marching Rule, their cargo cult or millenarian movement, where Sa'a villagers patrolled the beach after World War II looking for cigarette butts and boot footprints signaling the return of their American brothers in submarines who would carry them home to New York Heaven. This type of cargo cult became the conceptual basis for my Cargo Art—cargo being Pigeon English for manufactured goods. Through assemblage, I meld spirits from different worlds, the ones of nature and village life with those of industrial society, creating anthropomorphic spirit figures representing the power of the combination.

Children and I played together. They taught me Sa'a or Fanalei language, and their fantasy play and games. I detailed the ongoing revival of the Sa'a Village ritual tuna hunt by the Sa'a "bush people" in their elegantly narrow sewn-plank canoes inlaid in chambered nautilus. Ten miles up the coast in Fanalei Village, I swam in the ritual porpoise hunt of these Lau "saltwater people," who wrestle porpoises into canoes once they are herded into a cove. If I had done otherwise, it would have offended the porpoise spirits, ruined the hunt, and I'd have been ejected from the village. Recording everything I could in my daily journal, I began to understand that the idea of a "tradition" as fixed in stone was inaccurate. Rather, everything was constantly in flux, transforming.

Returning from the Solomons to live in my home village, the Upper Westside of New York City, I started to volunteer at the American Museum of Natural History. I'd skipped graduate school to fulfill my own Margaret Mead fantasy in the Solomons, but now it was clear that to become a curator and keep exploring while earning a living required advanced degrees. In 1974, Margaret Mead was still the museum's curator for Oceania, her permanent post until she died in 1978. Mead was an icon of social justice, speaking out for women's rights, minority rights, and for better ways to socialize children in our industrialized world with ever more isolated nuclear families. With all this in mind, here I was, standing in front of Margaret Mead, my hero, holding out my precious field journal.

I was calling on Margaret Mead high up in her turret office behind the scenes at the museum, floor-to-ceiling books and papers rising up around her powerful presence with spring light slanting in on her face, low behind her desk. Before her I placed the field journal detailing play with the children, their games and stories, and how they were socialized. My journal spoke of social change, including current village gossip—coconut news—about concern and confusion over Japanese fishing boats that seemed to be teaching their tuna to prefer to eat other hooks. I'd written it to keep my own voice company in the villages, to record what I was experiencing, to wonder at new ways, and because it just felt like the natural thing to do. Ethnography has always been my single greatest skill. Scanning through these pages, Mead proclaimed, "You have put the cart before the horse." She pulled her magic strings in her home anthropology department at Columbia University and got me accepted after all the formal deadlines. Mead changed my life.

My Solomon Islands field journal became the backbone of many graduate school papers. Professor Joan Vincent, my Columbia adviser, mentored me patiently in political economy, examining transnationals in the Third World—early work on examining what is currently called globalization. I aimed to return to find out how islanders in Sa'a and Fanalei were interacting with the new Japanese fishing joint venture with Taiyo Gyogyo—the largest fishing company in the world. It was a Solomons on the brink of 1978 independence from Britain.

With the generous help of the Rockefeller Foundation, I found myself back in that archipelago as a doctoral candidate, this time spending more time learning from adults. I was beginning a career as an ethnographer of coastal communities where people are interacting with global fisheries and industrial forces. I designed my fieldwork as multisited, spanning from Solomons villages of tuna or porpoise hunters, to the Japanese-Solomons joint venture tuna fishing bases, to the national and regional tuna fisheries levels, to the heart of Tokyo fisheries business and Rome's international fisheries research center at FAO—Food and Agriculture Organization.

DEVELOPING HUMAN/ENVIRONMENTAL RESEARCH

In 1984, I took my first job as a professor at the University of Miami's marine school, where I continue today. Having been chair for eighteen years, I built our division of Marine Affairs and Policy (MAF), hiring a team of faculty and developing a curriculum with specialty tracks. Marine Affairs—an interdisciplinary field by definition—blends anthropology, resource economics, fisheries and coastal management, aquaculture, and marine and coastal law. Marine Affairs has evolved into a national field examining human-environmental interactions, with programs in a handful of marine-oriented universities and an ardent student following among those who wish to bridge science and policy to conserve the world's oceans, manage fisheries and coral reefs, and help deflect the crisis of crumbling global environmental circumstances.

With the oceans in danger, sea level rising, and a high percentage of the world's poor living in coastal zones, this is critical work. Although many anthropologists tend to specialize in one region, I've chosen to specialize in coastal and marine human-environmental interactions, examining conflicts and developing comparisons and commonalities across regions, while building the interdisciplinary field of Marine Affairs and Policy. Exploring humans as part of the ocean and coastal environment, my fieldwork has become empathic, examining issues of social justice and social well-being.

During the 1980s, fellow anthropologist Edward LiPuma and I were a married team, collaborating together. We did several years of joint fieldwork on Spain's evolving industrial fishing fleets during their entrance into the European Union; went to the Solomons for the fiftieth anniversary of Guadalcanal; and examined the boom in Ecuadorian shrimp aquaculture that was decimating the mangroves. The havoc and devastation of this "Pink Gold Rush" was eye-opening for me, watching everyone flood in to cut down coastal mangroves to develop shrimp ponds for the cavernous U.S. export market. People were ripping up these critical nursery areas despite having imported the elaborate U.S. coastal zone management system. Only an epidemic in this monoculture squelched the rush into shrimp production.

This Ecuadorean fieldwork was seminal for me, as well as highly cited by coastal management during that 1980s boom and its demise. It was my first experience on the South American coast, and pointed me to the force of market booms. Throughout this book run the common threads of exploring fisheries booms, the management that follows in their wakes, and the crashes, including extreme competition and livelihood decisions in light of major El Niño events.

Comparing and contrasting important marine and coastal fieldwork themes from all my previous and ongoing research set the stage for research from Galapagos to Patagonia. These themes include land development and rising taxes pushing out fishermen in the Florida Keys; artisanal fishermen facing stock shortages because of industrial lobster fleets in Corn Island, Nicaragua; frequent deaths among Miskito Indians who dive dangerously to maximize income, and local processors refusing to fund their request to switch to traps on Corn Island, Nicaragua; the collapse of an indigenous reef management system, and how ready access to nylon nets and an urban cash market for reef fish cause reefs to be wiped out in Fanalei, Solomon Islands; the rise of women in fisheries management in Illa de Arousa, Galicia, Spain; and grassroots efforts to develop alternatives to overzealous fisheries in Isabela, Galapagos.

SERENDIPITY IN SOUTH AMERICAN COASTAL WORK AND FUNDING THANKS

Over twenty years later, I found myself hiking along cliffs near a fishing settlement in the very coastal villages where I'd first intended to do my Watson fieldwork. I burst out laughing, twirling and twirled by the stiff winds—how wonderfully bizarre! The serendipity of life's cycles of intention and desire hadn't dawned on me until that moment, long after writing and winning the multiple year NOAA grant that brought me to Chile.

Through fate, I had proposed work along the coast of South America without realizing the coincidence of choosing the same shores as I had originally proposed to the Watson Foundation after college. More than two decades and several field sites had elapsed when it became evident that the topic "human dimensions of global climate change" was a priority.

I found myself focusing on the identical Pacific coast region of South America where I had dreamed of sailing solo. Human dimensions of global climate change primarily examines how major El Niño events interact with human life, and grants for this topic were available. Livelihoods along the coast of Pacific South America were being deeply affected by these climate variations.

With this in mind, three of us formed an interdisciplinary team (marine anthropology, environmental law, and marine biology) and obtained our first NOAA grant for Chile. Our original ambitious field plan was to cover Ecuador, Chile, and Peru together, but it proved too much, so we more modestly proposed starting with Chile. Thus, in 1995, I found myself alone in northern Chile, writing the first field notes that would translate into this book, focusing on the relations between industrial and artisanal fisheries. As this grant drew to a close, we obtained another grant from NOAA.

Because this brand of fieldwork ripens over time, unearthing subtleties, I stretched my funding into the seven-year coastal odyssey related in this book, traveling the inland seas as far south as Patagonia, Chile, and up into the bays as far north as Paracas, Peru. I also received an anthropology grant from the National Science Foundation to continue my own ethnographic focus along the western shores of South America, comparing how different areas revealed different responses to climate variability.

Teaching ethnographic field courses as my métier, I have tremendous belief in fieldwork and experiential learning for coming to live at peace with cultural understanding and compassion. I began leading annual courses to Galapagos in 1999 and started fieldwork. This effort has evolved into a concrete attempt at environmental participatory management and social justice via the formation of Isabela Oceanographic Institute (IOI). Based in Puerto Villamil's Mission, the oldest structure on Isabela in the far western Galapagos, IOI is helping residents maintain control of their livelihood and grassroots tourism efforts within a fragile iconic environment.

Acknowledgments

Thanks flow for this long and detailed project linking many sites. Mentors, colleagues, students, editors, close friends in the field, and highly literate home friends and family have all contributed mightily over years of discussions, formulations, compilations, word games, editing, and encouragement. Many more deserve thanks, but here's my attempt at flowers for the living and the "living dead" who remain strong in heart.

Many thanks to my mentors. Paul Durrenberger encouraged me to recast the drier style full of numbers and dates by removing most of them in favor of descriptive stories with people speaking, voices breathing life into the ethnography and stimulating students to become engaged. Bonnie McCay inspired me with her eponymous work in maritime anthropology. James Greenberg, founder and editor of the *Journal for Political Ecology* and the Political Ecology Society (PESO), generously edited and commented on the importance of compassionate political ecology, supporting me in this approach. Gísli Pálsson kindly read and commented insightfully on the manuscript, and brought to bear his wide depth of scholarship on subjects ranging from maritime anthropology to phenomenology. He originally suggested I use "empathic" to capture this way of doing ethnography. Joan Vincent, as my adviser at Columbia University, trained me in political economy, examining transnationals in the Third World—early work on what is currently called globalization. Her voice urged me to turn my ethnography skills into a book that reflects the heart and science of fieldwork. Like so many others, I also wish to thank Margaret Mead, who literally set me on the path at Columbia University to formally become an anthropologist.

My former husband and close colleague, Edward LiPuma, kindly read the introduction. My fictive brother, Rich Mynick, liberally contributed wordsmithing. Sisters Lois and Janet Jensen shared their understanding of international development and a great ear for titles. Alaka Wali and Olga Gonzalez influenced my ethnography and spurred me on. Rod Neumann, without realizing it, continues to inspire me with his sensitive political ecology approach to ethnography.

Huge thanks go to Su Benaron for editing the manuscript with fine wit and insights. Boundless gratitude to Wendi Schnaufer, executive editor, Acquisitions, at AltaMira Press, for believing in this nonconventional empathic environmental ethnography. I'm enormously grateful to Doug Tyrrell of International Assets, who jumped in and solved some last-minute mapping problems with creativity and style.

Thanks go to my amazing RSMAS master's graduate students: those who shared fieldwork experiences with me in Paracas and in Iquique, Gletys Guardia-Montoya and Ricardo Gonzalez Hunt, Kenny Broad, and Juliet Gil; those who served as teaching assistants in my Galapagos fieldwork courses, Lila Asfour and Veronique Koch; those who coauthored articles on Paracas, Pisagua, and Chilean management areas, Lila Asfour, Yair Lichtenstein, Gletys Guardia-Montoya, Ricardo Gonzalez Hunt, Veronique Koch, and Mike Lemons. Mike Lemons deserves warm thanks for compiling my ethnography into the first version of the manuscript. Johann Besserer and I share a strong bond for cofounding Isabela Oceanographic Institute and carrying out a common cause for grassroots conservation, education, and geotourism. *Abrazos fuertes*.

I deeply thank my kin and fictive kin for kindness and humor through the whole process. Annie Laurie, evolved creative force, who is my best friend and sister, and Samuel Armistead, brother-in-law for over thirty years, legendary medieval Spanish folklorist and prodigious author, singing ballads and reciting poetry, they both are forever generous of time and spirit over this lifetime of ethnography. My mother, Alice, compassionate worker for social justice, stage actress turned social worker, supported my transformation into an anthropologist. My father, Stanley, artist of science, history, and fish, introduced me to the underwater world as a child. My fictive mother, Blanche Brown, NYU professor of Greek art history and Alexander the Great, was a role model, a vivid storyteller, and effusive inspiration. My fictive grandmother in Princeton, Jeanette Mirsky, rode in my heart as a force for cultural observation in anthropology and art. She was one of the early anthropology students at Columbia with Margaret Mead on Boas's knee.

Each field site holds many who have given me a sense of place and home, sharing their feelings and ideas. Among myriad Chilean friends who helped me, thanks to Wolfgang Stotz, who is my Coquimbo friend and coauthor on the article analyzing Chile's management area system; Ricardo Vicencio, and Maritza and Pepe in Iquique; Gloria and Ladiseo in Islas Galas, and Juan in Chiloe.

In Galapagos, thanks on Santa Cruz go to Jack Nelson and Matthias Espinosa; Jacqueline DeRoy, pioneer and silversmith of marine iguanas; Godfrey Merlin, dedicated conservationist; Rene Sangolquin, a great guide over the years; and his mother-in-law, Doña Anita Chapi, first colonist of Santa Rosa settlement. On Isabela, thanks start with the indomitable Sara Luz Ruiz Escobar, the compassionate and environmentally wise local director of IOI, and her husband, Fernando, ex-fisherman and IOI associate. Mireya and Manuco Caiser became my friends when I first arrived in Villamil, taking me snorkeling with sea lions and white-tip sharks, patiently showing me local life, from the *finca* to fisheries to tourism. Jose Gomez and his family have shared local knowledge and history, as well as their world-class coffee.

In addition to thanking him for coming up with the mnemonic CAGE (class, age, gender, ethnic identity) while attending my Galapagos field course, I thank Provost Tom LeBlanc for his foresight in bringing IOI into the University of Miami to create UGalapagos.

Great thanks go to both the National Oceanographic and Atmospheric Administration (NOAA) and the National Science Foundation (NSF), and their programs dedicated to understanding the human dimensions of climate change. Their generosity funded my field research from 1995 to 2001 in Chile and Peru, from Patagonia to Paracas.

Finally, I thank the Thomas J. Watson Foundation for initially awarding me their grant to carry out fieldwork along the Pacific coast of South America. Though I altered course to the Solomon Islands and then Spain, I ended up back in these Latino Pacific waters through serendipity.

Acronyms

BI	Bahía Independencia (Peru) Independence Bay
CEPAAT	Centro Profesional de Asesoría y Asistencia Técnica del Norte (Chile) Professional Advising and Technical Assistance Center of the North
CONAF	Corporación Nacional Forestal (Chile) National Forest Corporation
CONAPACH	Confederación Nacional de Pescadores Artesanales de Chile Chile's National Confederation of Artisanal Fishermen
CORFO	Corporación del Fomento de la Producción (Chile) Chilean Economic Development Agency
CRUB	Comisión Regional del Uso de Borde Costero (Chile, Region I) Regional Commission for Shoreline Management
DIGESA	Dirección General de Salud Ambiental—Ministerio de Salud (Perú) Environmental Health General Directorate—Ministry of Health
EEZ	Exclusive Economic Zone
ENSO	El Niño Southern Oscillation
ESPOL	Escuela Superior Politécnica del Litoral (Ecuador) Polytechnic University of Guayaquil
EU	European Union
GNP	Galapagos National Park
ICM	Integrated coastal management
IFOP	Instituto de Fomento Pesquero (Chile) Fisheries Development Institute

IMARPE	Instituto del Mar del Perú (Peru) Peruvian Marine Research Institute
INRENA	Instituto Nacional de Recursos Naturales (Peru) National Institute of Natural Resources
IOI	Isabela Oceanographic Institute
LME	Large Marine Ecosystem
MA	Area de Manejo (Chile) Management Area
MIPE	Ministerio de Pesquería (Peru) Ministry of Fisheries
NGO	Nongovernmental Organization
NOAA	National Oceanic and Atmospheric Administration
NSF	National Science Foundation
OMAI	Women's Organization for Artisanal Initiatives (Puerto Villamil co-op)
RNP	Reserva Nacional de Paracas (Peru) Paracas National Reserve
SERNAP/ SERNAPESCA	Servicio Nacional de Pesca (Chile) National Fisheries Service
TAC	Total allowable catch
UCN	Universidad Católica del Norte (Coquimbo, Chile)
UNESCO	United Nations Educational, Scientific and Cultural Organization
USAID	United States Agency for International Development
WWF	World Wildlife Fund for Nature
ZOFRI	Zona Franca de Iquique (Chile) Iquique's Free Trade Zone

Chapter One

Climates of Change and the Ways of Empathic Ethnography

3rd Fisherman: I marvel how the fishes live in the sea.
1st Fisherman: Why, as men do a-land: the great ones eat up the little ones.
—William Shakespeare, *Pericles*, act 2, sc. 1, l. 26–29

Coastal people and their sea lion counterparts epitomize sea change as they struggle with changing climates. The intricate and intuitive tales I tell from the Latino Pacific focus on coastal communities and how people cope with the continuum of chaotic changes that come with harvesting the ocean and living in increasingly desirable places where tourists and the wealthy want ocean property. Changes include climate variability in the form of extreme El Niño events that radically alter access to marine resources for good and for bad; migration of fishermen as a survival strategy; creation of tantalizingly lucrative export markets; concomitant fisheries booms with buyers' recruitment of fishermen to meet demand; governments' lagging policy responses to first-time booms; general oceanic decline of wild-caught fisheries; and governments' reluctance to trust and accept fishermen as partners in controlling marine resources.

Coastal communities along the Latino Coast of South America deal with periodic warmer waters and storms generated by El Niño events. Events cause providential fluctuations in the marine life people depend on. Environmental changes can trigger wild proliferation in a species, generating a market boom. But sea changes can also devastate a targeted species. Whether events bring feast or famine depends on the individual species and the unpredictable character of each event. People are eager to know whether the warmer water event will be intense, long, and onshore or offshore, but to date, climate models cannot provide this vital information.

This book tells six stories that emerge from seventeen years of multisited ethnography in Chile and the Galapagos Islands (figure 1.1), and draws on work done in Peru during the same time. My fieldwork along the Latino Pacific Coast was conducted from 1995 through 2001, and in Galapagos from 2001 continuing through 2012. Throughout the book run common currents of exploring fisheries booms, the management decisions that follow in their wake (usually too late to accomplish their goals), and the subsequent crashes, the survival strategies of coastal people, and the extreme competition that arises among various interests in fisheries, tourism, and conservation.

As evidenced in this book, human/environmental interactions during the Anthropocene cry out for global environmental change research (Pálsson et al. 2012). Pollution, global warming, population growth, and overfishing put enormous pressures on ocean ecosystems and fish communities. In these stories, we see how the ocean is clearly a key source of food, income, and social well-being, as it is for coastal communities around the globe.

CLIMATES OF CHANGE

The National Oceanic and Atmospheric Administration (NOAA) and the National Science Foundation (NSF) run human dimensions of climate change research programs that will deepen insight into fisheries decision making and management. I am grateful for their generous funding of the fieldwork for this book, which aims to understand fisheries decision making and management processes through a suite of ethnographies that follow Latino coastal people as they interact with climates of change. We observe six different groups of fishing people who navigate diverse marine outcomes through boom and bust fisheries cycles caused by the last radical El Niño event in 1997–1998. The ethnographies share overlapping themes of competition against industrial or outsider fishermen; the lure of beachfront real estate and the lust of coastal developers; evasion of top-down management and temptations to smuggle; and the need to reorganize and lobby for Management Areas and aquaculture. Ultimately we see people politically and economically angling for control over marine resources.

From the stories, a pattern emerges of fishing people watching to see the exact marine ecosystem uncertainties triggered by El Niño events before they act. Through fieldwork in multiple sites, I observe how variations in environmental forces forge different outcomes. No two events play out the same way in terms of what really counts to the coastal people: the duration of the event, its intensity, and whether it is near shore or offshore. Environmental change is on a trajectory to produce more extreme El Niño events. Coastal people

Figure 1.1. Latino Pacific map. Adapted from base map from https://www.cia.gov/library/publications/the-world-factbook/.

interact with such environmental events set within the context of short-term, shifting political and economic climates.

These overlapping changes determine critical access to contested marine resources. Strange bedfellows emerge among the three key umbrella interest groups—fisheries, conservation, and tourism. Interest groups (stakeholder groups in the political parlance of development) have members who by definition share certain beliefs and practices. Unlikely alliances arise as players strategize and struggle to control marine resources. In this light, my mantra has become constant climates of change: political, economic, social, and environmental.

El Niño as Environmental Climate Change

El Niño—short for El Niño Southern Oscillation (ENSO)—is a cyclical climate phenomenon that has drastic social and ecological repercussions worldwide. NOAA (2009a and 2009b) has created lucid websites describing these processes, which I plumbed for information. ENSO represents the largest interannual variation in the global oceanic/atmospheric system. The atmospheric contribution to ENSO is the Southern Oscillation. The Southern Oscillation refers to the cyclic increase and decrease of the Southern Hemisphere trade winds caused by varying atmospheric pressures in the South Pacific. During an ENSO event, the trade winds weaken or reverse, and the warm surface water in the western Pacific Ocean travels in a wave to the east causing the thermocline to deepen along the South American continent.

The oceanographic contribution to ENSO is El Niño, which refers to an increase in surface sea temperatures in the central and eastern tropical Pacific Ocean, especially off the South American coast. Positive temperature anomalies may be as great as ten degrees. The warming of the surface waters depresses the thermocline and impedes the upwelling of nutrient-rich colder waters. This results in a decrease in the productivity of surface waters and a reduced food base for zooplankton. Through these processes, ENSO events on land induce floods, droughts, and other climatic perturbations in locations around the world. In the oceans, there can be conspicuously large changes in the abundance and availability of important living marine resources in the eastern Pacific Ocean. Again, the changes that matter most to the fishermen are the event's intensity, duration, and location on- or offshore.

Empathic Ethnography as a Tool for Discovery

We can observe others as

> fellow wayfarer[s], struggling with [their] burdens, wanting happiness and dignity. Beneath the fears and needs, the aggression and pain, whoever we encounter is a being who, like us, has the tremendous potential for understanding and compassion, whose goodness is there to be touched.
>
> —Jack Kornfield, *The Wise Heart: A Guide to the Universal Teachings of Buddhist Psychology* (2008)

This section provides the background on my approach to ethnographic research, exploring the field methods I use and the theoretical approaches that guide me. The ethnography follows the lines of political ecology in human geography (Neumann 2005) and anthropology (Biersack and Greenberg 2006). One thinks in terms of interrelationships at multiple political levels, putting everything into historical perspective, while understanding that a community is hierarchical and not homogeneous. The nature of agency intrigues me. Political ecology tends toward structural and historical explanations, opening questions of change. My work looks at how vectors of change located in the environment (e.g., El Niño events) affect fishermen. Through ethnography, I simultaneously analyze fishing people interacting with these chaotic changes in marine systems and short-term or sudden shifts in political and economic relations, thus raising the question of agency.

I am a believer in poetic, descriptive ethnography. To this end, ethnographic sketches are woven throughout the book to lend the feeling tone of people's existence, offering sentient notes. I want these sketches to convey insights people reach about their own situation and conflicts; the reasoning and rationales behind their decision making. I want fishermen's tacit knowledge and expertise to resonate with those of us viewing the world from other angles, to open up our perceptions; to make them more inclusive, ultimately to widen our comprehension of how humans cope in changing environments. Thus, these sea lion stories are interwoven with the past, dynamic and chaotic in a way that leaves little room for dispassionate deconstruction or loss of felt reality. This book is my attempt at interpreting issues laden with emotion and deserving of words that evoke feelings.

A dearth of story genre has led me to write ethnography in a way that keeps individual voices telling their stories in the foreground. This reveals the lives of coastal people through personal narratives, focusing on the experience of fieldwork, social relationships, difficulties, and simply the uncertainties of doing ethnography itself. In parallel to the academic enterprise of parsing people into interest groups to map competing perspectives, I highlight the storytelling approach to emphasize the empathy that arises during fieldwork. Thus, much of what follows is in the form of conversation, a structure that retains the personal voice in the exchange of information and echoes the give and take of the ethnography process.

Empathic ethnography, by its nature, must be highly personal; an exploration of how anthropologists and geographers interact, empathize, and learn from the people we live with and write about. This series of ethnographies offers an account of doing fieldwork in multiple sites, observing people relating among themselves, struggling and negotiating, muddling through and adapting to change. My mission includes enticing and guiding students into the field for hands-on research, and preparing them to become future travelers with wide-open eyes, new friends, and a daily observational journal.

A powerful, comparable "many-stranded conversations" approach to ethnography is taken in Gudeman and Rivera's *From Car to House* (1995). However, my route is continental rather than local, more by boat than car, and the time scale is expanded by years. This is a personal story, a conversation with fishing communities, with minimal referencing to the literature, not an objective report. The conversations and stories resonate with the empathic core of my book's ethnography.

This book illustrates how to do fieldwork by example. Focusing the chapters on ethnography, I have left the connections to academic conversations implicit. Numerous vignettes, anecdotes, opinions, personal letters, and instances of a more poetic prose are sprinkled throughout the book to bring out the excitement and flavors of fieldwork, living among people from unique and distinctive cultures.

By peppering the book with stories, I'm flavoring the landscapes with the ardor innate to the primary ethnographic venture. Each vignette is a collage, speaking to the excitement of fieldwork. It's a narrative about specific places and how we as anthropologists and geographers come to know what we know. What has been called the art of ethnography, commonly left a mystery, is divulged through discovering changes.

Here, for the reader's benefit, I present some of my underlying assumptions. For the qualitative style of ethnography, I've excised most of the flood of numbers and meticulous details that may obscure the larger picture. These are the drily factual nuts and bolts that we anthropologists and geographers diligently collect to help tell us what is happening. Paring these down to the bare minimum, so the forest isn't hidden by trees, I offer the distilled essence with the essential details.

In contrast to the usual academic studies, this book strives to demonstrate how ethnography can embody empathy. Thus, this book disseminates hypotheses and theories, suppositions, and possible changes as solutions through a language that directly mimics the processes from which these ideas arise in the first place. This language consists not of strings of the latest polysyllabic fashion or buzz, but rather of strings of personal accounts rooted in concentrated observation, rumination, and—more often than is presumed—an unspoken instinct or awareness.

Empathic ethnography, as I practice it, is the underlying framework that expresses itself without need of explicit explanations. In each chapter, readers can imagine themselves meeting people at ease in this openhearted way. They can look through my eye to appreciate the serendipity of the moment, and get a sense of the spontaneous strength with which fieldwork interactions may unfold.

Empathic Ethnography in the Field

This section discusses the methods behind the ethnography and offers the tools for this approach to fieldwork. These tools can be used in a variety of settings to garner a deep understanding of interest groups and their interactions. Though not mentioned explicitly in the ethnographies, these are the tools that I use and that I teach to my students to prepare them for the fieldwork experience.

Beyond the de rigueur randomly selected numerous interviews wherever I have carried out fieldwork, my route to obtaining a fish-eye picture invariably includes intense relationships with a handful of individuals—a number are recognized leaders, whereas others are on the fringes but still have visionary appeal. For me, the fieldwork process entails us gravitating toward one another in an openhearted way. It's always fascinating to discover the official and unofficial leaders, as well as the creative souls circling on the periphery. Many feel circumscribed within their customary boundaries and curious about the outside, thus curious about me, who has suddenly materialized in their world. We seem to gravitate toward each other, mutually intrigued. My social network grows geometrically.

Throughout fieldwork, I'm routinely spending time within coastal communities, whether deep in interviews or helping to mend gear, watching the offloading and marketing, gossiping with elders, hearing histories, listening to views on today's top thorny issues for folks, playing with the kids, helping to cook, and overall enjoying getting familiar with local life and its hierarchies. Initial conversations, a helpful attitude, and mutual kindnesses foster bonds. I've found myself sitting in a circle of older girls and women knitting llama wool in the *altiplano*. Asking how to locate grazing llamas, moments later we're making plaintive llama cries back and forth, as I try to echo their llama communication. Word spreads rapidly about this entertained/entertaining newcomer (me), and what we've been doing together. The inquisitive approach! We exchange stories of an industrialized world and theirs. Doors swing open in a domino effect.

I also attend group meetings—situational analysis in the language of grantsmanship—and increasingly discern which individuals are the driving forces and initiative takers in different situations. These people are consistently the go-getters, the movers and the transformers, often renaissance people. They want to fathom foreign places and grasp what other systems exist. We each want to connect to the other's reality. Sometimes if I'm lucky, I can be useful by linking them to grants, scholarships, or outside agencies. It's a sharing, a process of mutual storytelling-as-learning that lends insights and facilitates probing the depths through further discussion.

In small communities, it's palpable who has specific vested interests and how they are potential fonts of wisdom. Meanwhile, in large or even urban places with so much occurring simultaneously, it's more challenging to unravel. But wherever fieldwork has drawn me, it's a mutual process of being drawn toward those thinking about or involved in social transformation.

Participant Observation

Ethnographic fieldwork revolves around participant observation, anthropology's core contribution to social science. Formal elements combine with intuitive detective work. As a mnemonic device when teaching ethnography, I've dubbed this way of learning "PO"—the acronym for participant observation—in honor of Edgar Allan. PO is the door to joining in discussions, daily tasks, meals, games, and meetings in the great attempt at feeling others' passions and problems from the inside; how they experience their lives, their views, choices, and sense of purpose. I'm constantly reciprocating with my own thoughts, emotions, and personal stories. Friendship, trust, and confidence can grow through such sharing. This is fieldwork in action.

Conducting in-depth, open-ended life and work histories is a necessary complement to PO. Through participant observation we watch people's actual decisions and interactions, and compare these to what people say they do, which is often their idealized behavior. As people describe in detail their families, communities, and jobs, we follow their lead into free associations and what may seem at first to be sidebars. These interviews are free-flowing conversations that are gently guided toward exploring the tangents hinted at by the teller. The storyteller and the listener first establish rapport, a human kinship, a bond of fellowship. Though systematic data collection is essential to my endeavors, the real strength of my ethnographic method comes from self-awareness allowing such empathic connection between people. The ethnographer achieves this level of genuine listening through a viscerally felt level of concentration, recognized at an intuitive level. Intuition includes reading eyes and facial muscles, postures, intonations, and other such expressive clues through being thoroughly connected and engrossed by the storyteller.

Work histories unearth how the twists and turns of fate lead someone along their livelihood path. We discover how someone changes jobs from their first one to their current position, why they think they shifted positions, and about their family and other social networks who helped in this process. We learn who they work with directly and indirectly, from family members to friends or contacts. We find out their formal and informal education and sources of practical knowledge. We come to understand patterns of migration within circles of family and friends and where their money flows (e.g., remittances home, common investments). A moving picture comes into focus. It's set in historical perspective, with details such as who are allies and rivals;

how access to resources is shifting; and who controls the equipment and infrastructure necessary for work.

Interviewing the same person repeatedly, perhaps in different settings, yields varying emphases if one follows the threads carefully. Life histories overlap with work histories. A life history starts with a person telling their story, usually describing where they were born and raised. Beginning with their parents and siblings, their story fans out to include as many family members, in-laws, and fictive kin as possible over generations, complete with their viewpoints and a wealth of detail about daily life.

Individuals' concepts emerge, such as what makes a good parent and child; member of a community; member of a religious, political, work-related, or other institutional group; and steward of nature or conscientious fisherman. We see the tellers' awareness and sense of obligations, interdependencies and relationships, and how they interact across generations as society changes.

We ponder why a person may hold contradictory or controversial viewpoints compared to others interviewed or policies on paper. We are not seeking objective truth, but rather contrasting their reality, concepts/ideals/beliefs, and practices/behavior/actions with those of others. We all fool ourselves about how we really act. This is the foundation of another basic wisdom from ethnography: concepts inform and influence practices but rarely match them.

Interest Groups—Mapping Perspectives

Having considered which interest groups at all political levels are relevant to the project, we organize individuals into these groups. Interest group members by definition share PICs. We match commonly held Concepts derived from Institutions (e.g., family, religion, political party, school, clubs), inspiring Practices. Through life and work histories, we also look for patterns coming from a person's "CAGE"—the mnemonic standing for: Capital (e.g., class based on various types of capital such as financial, educational, cultural, and family connections); Age (generation); Gender; and Ethnic identity. People take their personal CAGE as critical reality.

Capital brings access to multifaceted sources of power and resources. It comes in various forms such as land wealth; ascribed family stature, lineage status, and family social networks; financial capital; formal educational capital; skills and local knowledge; and cultural and artistic capital. Class—the word popularly bandied about—is far less nuanced than thinking in terms of different types of capital. From culture to culture, one's class is considered more or less malleable and does depend upon different types of capital.

Ethnic identity—self-identity based on genealogy as well as such shared features as language, style of child rearing, familiar food and spices, family traditions, and spiritual beliefs—is far more multifaceted than the socially simplified American category of race. Ethnic identity plumbs self-identity, understanding shared perspectives and the powers of that group as viewed by themselves and others. It is fine-tuned and detailed, often revealing the complex choice of ethnic backgrounds important to an individual in their society.

Mapping the perspectives of the various interest groups—scientists, managers, and developers included—is a formidable technique for determining the contour of the entire picture. It opens pathways for communication among groups, and thus promotes change. Fieldworkers as well as members of the groups become acutely aware of the stakes, competing viewpoints, and diverse logic. We map what each group thinks it stands to benefit or lose, who are its allies or rivals for resources.

Mapping competing perspectives makes mistrust transparent among certain groups and reveals how they think management or a potential project will alter access to resources. It suggests—given internal logic—who might decide to support and follow a particular plan rather than ignore, undermine, and/or resist it. In development or management projects, mapping can lead authorities to more nuanced plans beyond simply recruiting, hiring, training, or entrusting resources to the most conspicuous group leaders who are assumed to be in control. More can be accomplished at the grassroots level by carefully considering hierarchical relationships at each level to avoid group competition, uncertainty, and lack of trust. For example, the most powerless in a community who are asked from above to invest sweat equity in a project often doubt they will be included in reaping future benefits. Thus, mapping is a powerful tool, whether a project advocates incorporating locals in a management process (participatory management), or passing down regulations and policies from central government (top-down management).

A Multilevel Approach

The beauty of the fieldwork demonstrated in this book is its multilevel approach, which in turn calls for multisited ethnography. By situating myself at several sites, fieldwork always gains fresh angles on common questions. I develop a real physical presence and explicitly weave sites together to generate the whole cloth of ethnography. This provides depth of field, a sharp picture, and a sense of overlapping processes. The multisited approach with its powers (Holmes and Marcus 2008, Marcus 1995) advances ethnography and needs to be widely adopted.

The keys I've developed over the years for fieldwork and teaching ethnography in the field have always included the multisited approach. From different cultural settings I've been able to explore multilevel political forces and gain a historical perspective. It is critical to comprehend that communities are heterogeneous, comprised of competing groups. The culmination is not the mere plotting of interrelationships among interest groups with their ever-changing, overlapping, and conflicting control of resources. An overly analytical eye risks doing violence to its subject, rendering whole lives into categories, bits, and over-fine theoretical distinctions. It can miss the 3-D unifying heart of the matter. Perforce, my intention as an ethnographer includes capturing the range and flux of meanings and emotions that infuse us moment to moment.

The ethnographies presented can teach the reader to recognize common themes and prevalent patterns in coastal cultures. All discuss potential shifting patterns in power and access to resources, given climates of change and limited household time and resources. As a set, these ethnographies can also provide insight for conservationists, consultants, or managers involved in designing and evaluating coastal management projects. They can be used to critique coastal real estate and tourism development, or guide toward increased social well-being.

For those interested in ethnographic research, this book provides an understanding of how to carry out this kind of research, including a sense of what the field might be like, the kinds of questions to ask, and a framework to understand interest groups. For those teaching about coastal cultures, each of the chapters can be used as evidence to allow students to interpret events, draw conclusions, and support theories. One reason that much of this book is in the form of conversation is to retain something of the interview process. Though these are not transcripts, students can learn about interviewing from the style and content of the dialogue.

For those interested in community grassroots development, particularly fishery and coastal development, this book offers insights into incorporating fishermen into the management process, allowing them to take ownership for their own ecosystem and simultaneously allowing them to see for themselves that their livelihood is not jeopardized by management.

For future managers, leaders, and NGO workers, this book provides ideas for designing a management environment that is acceptable to fishermen and managers alike. These ethnographies reveal definitive questions for critiquing and improving projects and incorporating a way to learn to predict how projects might affect local-, national-, and international-level interactions, and how people interact with the climates of change.

SEA LION STORIES

Sea lion-human interactions throughout the book epitomize sea change. Fishermen can see themselves in sea lions; they watch these charismatic marine mammals who must also cope with the environmental change of El Niño–induced scarcity or plenty. Sharing the water world with sea lions, coastal people recognize these playful graceful swimmers as their metaphoric selves, whether as mascots or rivals, as they cope with artisanal competition with industrial fleets, tourism, or market demands influencing regulations. Sea lion/human relationships underscore their deep imbrication, not an opposition of nature and culture.

"Sea lions" signify fishermen as well as their marine mammal compatriots who share the evolving scenes throughout the book. Sea lion/human interactions epitomize the perpetual sea changes in environmental, economic, political, or social realms. They underscore environmental tensions and difficulties of management, despite existing policies or international funding for conservation, and the ongoing shifting struggles among tourism, fisheries, and conservation interests.

In Galapagos and in Chile, with the crash of fisheries booms, fishermen earn alternative income by taking tourists to photograph or snorkel in *loberias*—sea lion nurseries (figure 1.2). Fishermen give harbor tours that include ogling sea lions draped aesthetically over their boats at mooring. Sea lions even picturesquely snooze among the strings of plastic bottles circling gunnels despite attempts to scare them off. Sea lions' accumulated weight will sink a boat with its precious motor. As an attempted solution, fishermen build alternative floating platforms solely for sea lion basking.

Fishermen in the inland seas of Patagonia feed juvenile *merluza* that the market rejects to the gathered sea lions. Sea lions play toss and catch among themselves around the buyers' boats in Islas Galas harbor, feasting on juveniles tossed overboard because they fall below the policy size restrictions at the industrial processing plants. The sea lions, positioned between the buyers' boats and the landing data devour the juvenile part of the catch, skewing the landing data so they "prove" a dearth of juveniles. Yet stock assessors and managers persist in taking landing data as in-the-water reality for the inland seas. The decrease in juveniles is then used as scientific justification for cutting artisanal fishing quotas.

At the northern Chilean industrial fisheries port of Iquique, sea lions at the artisanal wharf have learned to rely on fish-cleaning scraps. Small-scale fishermen bemoan how the industrial fleet has wiped out their mutual supply of fish—along with the big bull sea lion's pride they have their own pain over lost pride.

Living at the sea edges of Chile's northern desert in relative isolation from the law—including those protecting sea lions—a few people still enjoy grilled sea lion meat and drink blubber oil for virility, a sympathetic folk aphrodisiac. But during the starving years of a major El Niño event, men stand slump-shouldered by corpses of great bulls rotting on the beaches, ruing their shared hunger and suffering.

Loss of a prized abalone-like shellfish and other marine staples and the inordinately rough seas that limit fishing time at sea are tough. Fishermen see

Figure 1.2. Sea lions awaiting fish scraps at Caleta Riquelme wharf, Iquique. Courtesy of the author.

themselves in the sea lions, weathering these fickle outcomes of an El Niño event. Simultaneously, they speak of vying with sea lions for limited fish. They know how a hungry sea lion will steal speared fish literally out of hand, or grab into a net bag held at a diver's waist.

Looking at human/environmental interactions as they unfold moment to moment, I trace how they evolve in response to macro changes. Call it the practice of focusing on the here and now, considering fishermen's intentions and immediate interactions with changing conditions. Embracing a seemingly vast context, my lens views human sea lions as part of the marine environment.

NAVIGATING THE BOOK

The open architecture of this book allows the reader to wander among the six ethnographies to watch how events in different places can have divergent outcomes. This framework is referred to in architecture as *design build.* It thrives on the hands-on, evolving design process when the architect/ethnographer builds on site, observing the project unfolding moment to moment, steering to take advantage of serendipity and new interrelationships, planning the next step in response to immediate events in the field.

Any of the ethnographies may be taken as separate cases. All cases compare how people adapt to changes, despite similar backdrops of macro-political economic forces and the serendipity of environmental climate change in the form of El Niño.

To transition between ethnographic sites I have inserted letters that are taken from my personal correspondence from the field. The art of letter writing from the field is something I've always loved. It's the chance to personalize observations, merging gut experiences with free association and word play. The letters throughout the book are glimpses of life in the field originally shared with dear friends. I have taken out most of the interpersonal literate graffiti, keeping the anthropologist's eye. I have taken the liberty of cutting and pasting real content to supplement the ethnography and so that the letters segue into the next field site.

Ordering the ethnographic chapters, I begin with the place of quintessential environmental consciousness, the Galapagos. Since Darwin, this archipelago has provided global inspiration for conservation of rare and fragile environments, beginning with understanding of origins and evolution. From Galapagos, I drop down to Patagonia to follow the gyre of the nutrient-rich Humboldt Current, which gives the environment its character and extremely prolific marine life. Then I proceed up the spine of Chile's glacial south islands; along the central bays; into the extreme desert north coast (see figure 1.1). The following is the organizational layout of the six fieldwork chapters:

Chapter 2, "Wild West Galapagos: Isabela's Fishing Community Angles for Tourism Pie in the Land of Darwin," takes place in Ecuador's Galapagos National Park, best known as Galapagos (see figure 1.1). On the far western island of Isabela, in the boom-bust fishing community of Puerto Villamil, locals are angling for a share of the international tourism pie. Fieldwork chronicles their shift over the last decade from an isolated, small fishing community to one swept into the international tourism market. They have gone from an independent-minded community with a strong sense of pride and social well-being—despite the frequent deaths of sea cucumber divers—to a place whose boom fisheries crashed, leaving them with dreams of renewed diving but having to survive via tourism. They are transitioning to a broad-based community infrastructure that's an appealing destination. However, the Park agreed to allow the industrial-scale yacht tours to inundate Villamil, using it as one more tour site in the archipelago, the only human colony on the list.

Chapter 3, "Rain and Tears in the Fishing Camps: Merluza Boom to Bust in Patagonia," is set around Islas Galas in the inland sea of Chilean Patagonia's Regions X through XII (figure 1.3 and see figure 1.1). Farmers, other types of fishermen, and the rampant unemployed were recruited to fish these southern seas for a new export commodity, southern hake—*merluza*. The original rough fishing camps transformed into communities as women and children joined the fishermen. Fieldwork examines Chile's Pinochet-era politics and economics. It explores artisanal and industrial *merluza* rivalries and the power of markets to create and determine fisheries booms.

Chapter 4, "The Dance of Macha Divers and Waders around El Niño's Flood," and chapter 5, "Puerto Aldea: Caleta Life in Navy War-Game Waters," take place in Central Chile's Coquimbo area in Region IV (see figures 1.1 and 1.3). The two great bays of Coquimbo and Tongoy experienced opposite fisheries outcomes from the mighty 1997–1998 El Niño event. Competing shellfish divers and waders in both bays were compelled to migrate for survival. These chapters consider Chile's model Management Area system. Although fishermen formed associations and maintained mandatory self-management so they could apply for a Management Area, the government and Navy have resisted granting fishermen their promised marine tenure and management rights. This undermines fishermen's participation in management far more than the natural destruction from El Niño.

Chapter 6, "The Goliath in Iquique: Industrial Fisheries," is set in Iquique, Region I's historical industrial fishing port (see figures 1.1 and 1.3). Chile's powerful fishmeal and fish oil industries have evolved in relation to the great mining industries during the prime periods of economic and political transition. Among Iquique's industrial fisheries labor leaders, the political axiom on the stock seesaw was "President Allende and anchovies crashed when General Pinochet and sardines rose." Fieldwork examines how strategies for fleet and shore labor were altered in the face of unpredictable El Niño–induced stock swings.

Chapter 7, "Sea Wolves: Artisanal Fishermen of the Rural and Urban Caletas," concerns Iquique's artisanal fishermen (see figures 1.1 and 1.3). Urban fishermen and their rural counterparts, who first settled fishing camps south of the city, deal with the giant industrial fleet illegally trolling inside their coastal zone waters. Rival rural groups fight over marine rights and rob each other at night over Chile's first Management Area and the neighboring potential Management Area. Fieldwork compares rural and urban cohorts and ponders what community factors might lead to better management-area compliance. It further investigates artisanal-industrial controversies and the effect of government's antimigration policy on settlement strategies.

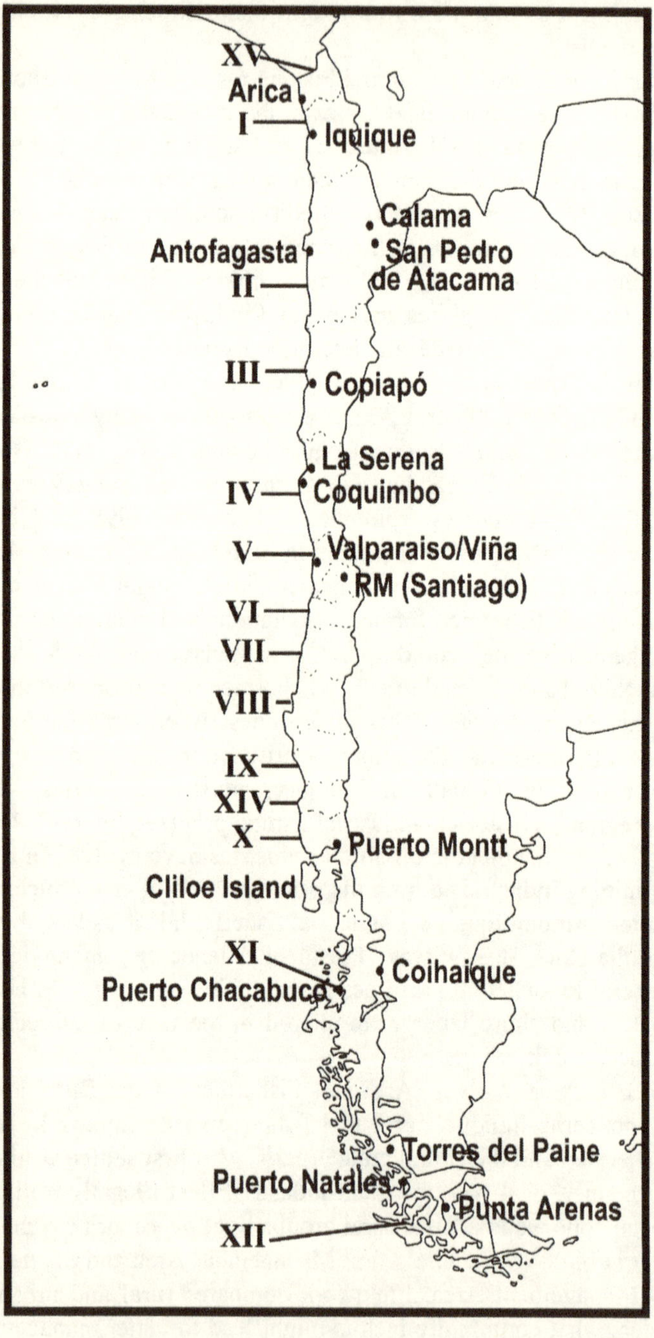

Figure 1.3. Regions of Chile map. Courtesy of Doug Tyrell. Adapted from base map from https://english.freemap.jp/.

Chapter 8, "Conclusions: Coastal Ethnography Given the Eddies of Climate Change," briefly sums up the book, highlighting its themes and patterns. It also examines my personal applied grassroots work as cofounder of the Isabela Oceanographic Institute Foundation (IOI) based in Puerto Villamil, in the Galapagos. At the community's request, IOI supports the community's desire for the skills to compete in the tourism sector and for projects that improve their quality of life and social well-being.

I hope readers will feel free to explore the fieldwork and storytelling here in a way that allows them to unearth their own conclusions, and compare these situations to others that arise in their own projects or travels. I have adopted Karen Brown's (2001) approach—offering the storys' fabric as a basis from which each reader can project and interpret themes and theorize. As new academic trends inevitably surface, the reader can reflect on this material as a springboard for further interpretations.

Last, I offer two brief notes on terminology and characters used throughout the book. The politically correct trend in development and ethnographic literature substitutes "fisher" to gain gender neutrality. In the Latino East Pacific, however, finding women fishing from boats is wildly scarce. Therefore, throughout this book, I freely and intentionally say "fisherman" instead of unnecessarily seeking gender neutrality with the currently popular "fisher."

Characters in the stories are often composites, authentic in flavor without revealing identity. However, a few of the main characters have agreed to let me use their names and tell their stories as best I can.

Chapter Two

Wild West Galapagos

Isabela's Fishing Community Angles for Tourism Pie in the Land of Darwin

> "The time has come," the Walrus said,
> "To talk of many things:
> Of shoes—and ships—and sealing-wax—
> Of cabbages—and kings—
> And why the sea is boiling hot—
> And whether pigs have wings."
> —Lewis Carroll "The Walrus and the Carpenter" in *Through the Looking-Glass and What Alice Found There* (1872)

REACHING TOWARD DEATH

The brown, nubby sausage-shaped, foot-long marine animal, *Isostichopus fuscus* is a treasured Asian market species of pepino that inches across the Galapagos seafloor on tiny suction feet, seemingly sitting still for the collecting. Pepino—sea cucumber or *beche-de-mer*—has been craved by the Asian market for generations and consequently harvested out of other seas. The pepino's slimy texture is enjoyed as a delicacy for its health benefits in Asia, but people from other cultures usually don't cotton to it. Discovered in Galapagos, it suddenly found itself the target of a thriving fishery, lying like a dollar on the sea bottom where divers collect it purely for the cash reward.

Max's captain cuts short their four-day pepino dive trip to rush Max back to Puerto Villamil. On the southeastern shore of Isabela, Villamil is the only town in the western frontier of Galapagos. Max, one of the young men who have moved to Villamil for pepino, grimaces in extreme pain. He has been "bent"—a dive injury caused by gas bubbles accumulating in joints from lack of proper decompression time when surfacing. Other fishermen cradle him to the local clinic where he's warned never to hazard dive again. Being bent once greatly increases the chances of dying the second time.

Julia, his wife, child still on hip, hair swept up off her round face, takes it as a portent and makes him swear he'll give up the allure of pepinos. She recruits her boss to reinforce the message. Julia works at Casa Marita, the first and most lovely beachfront hotel in Villamil. It's owned and run by Marita, who is an artistic, highly cultured Peruvian, and her Italian husband. When Marita listens to Julia's description of Max's dive accident, she calls him in and lectures him on family duty, offering to compensate the loss of diving with some work around the hotel if Max promises to forego pepino.

Yet on a weekend when Marita is away, Max announces to Julia that he's going on another trip to the best pepino grounds on the west coast of Isabela. They are headed to Canal de Bolivar—Galapagos's famous pepino nursery area. Julia begs him not to go. Max insists he's only needed to stay inside the boat running the hookah rig. Marita isn't there to weigh in.

Within a few days, Max's captain is chugging back into harbor bearing the tall young body of Max, still alive but writhing. The doctor dashes down to the dock bearing the clinic's meager supply of pure oxygen. He hopes to reverse the agony and stave off the onset of death. Watching him administer oxygen with the mask, locals mutter among themselves, spreading their deduction that the lack of oxygen must be the cause of the bends.

The field of dive chemistry and its invisible effects are neither widely understood nor believed. As the wharf crowd thickens around the victim, people shout hotly at each other how best to rush Max to Puerto Ayora, on the central island of Santa Cruz, where the only decompression chamber and dive specialist are located. It's an hour and a half by the fastest craft. Minutes matter. Though the Park/Navy possess such a boat, the authorizing official is not in town. For over an hour, arguments rage back and forth whether or not to requisition the boat without permission. Eventually, they load Max onboard with the doctor and Julia to begin the race across the ocean. Max almost survives to the waiting chamber. Without the dockside delay, Max might have received twenty minutes treatment. Julia is distraught in the arms of the doctor. He attempts to comfort her, saying Max had no hope even with the chamber.

Max's Ecuadorean family "on the continent"—as opposed to the Galapagos—is duly alerted. A small delegation makes their way to Isabela. They hadn't seen their Max for more than a decade, but they want to bring his body home for burial. Max has rooted himself in Villamil, having married twice, with kids by the first wife and by Julia. Max's relatives confront Julia when she says she wishes to keep her beloved's body. They then threaten to accuse her of murder unless she relinquishes Max, claiming she must have driven him to dive for money. The relatives start canvassing the whole tiny fishing community of Villamil, collecting donations to fly the corpse out in his coffin by renting a small plane. Villamil fishermen have a co-op that steps forward with money for the relatives, eager to have the case closed to minimize negative publicity on the deadly nature of diving for pepinos.

Around town, Julia is making it plain in her grief that she plans to hold Max close by and bury him with her large extended family in the coastal cemetery. Max's relatives approach me, too. I've been watching emotions seethe and want to support Julia. They say Julia wishes to give them Max. "I'll give her my donation to do what's best," I say.

The fight over the body revolves around keeping Max's spirit close. In Latino cultures, families celebrate Day of the Dead on the first of November in the cemetery. The living clean and redecorate the graves of their "living dead" and hold reunions, picnicking and maybe dancing to bands that circulate among the families. One can also commune with the spirit at any time by visiting the grave.

It's the ultimate moment; enough money has been found, and the rented twin prop whirs on the airstrip, while winds sweep over the freshly dug empty grave. Marita holds Julia, who has come to say goodbye, but they are helpless to halt departure. Gusts tangle hair across their faces.

Invisible Specter of Death

The trail of widows and mourning families left behind in Villamil each pepino season has been the shadow side of boom-time tales of wealth. Max died the year after the decade-long pepino boom that claimed so many other divers' lives as they harvested forty million individual pepinos. The specter of death continues to haunt Villamil women. How ironic that outsider relatives threaten widowed Julia with murder charges for pressuring her man to dive. It's women who usually plead caution and openly worry about the life-and-death trade-offs of pepino diving. At the opening of every pepino season, each woman is scared by the lottery-like nature of diving. Who will be widowed next or lose a dear brother or son, father, uncle, or cousin?

Divers themselves rarely verbalize or acknowledge this fear, like many men engaged in high-risk professions. Most divers don't bother to learn about cleaning the compressor filter so that the hookah rig provides a safe air supply. They disregard scientific dive chemistry with its invisible gas bubbles and continue to bounce dive—going up and down, up and down—diving deeper without sufficient surface time to off-gas the dangerous bubbles that fizz up like a shaken soda.

Pepino divers in Isabela, in spite of dive chemistry information, have kept pushing their bodies under physical pressure at greater depths to collect one more pepino. Though Galapagos divers do not attribute death to curses, the invisible science does not trump the financial lure.

"Pepino would be lying so close, just out of arm's reach, like dollar bills," says Julian, a sleek diver with half-hooded eyes who's fixing his gaze at the water. We're by the wharf, watching as early morning penguins herd bait against the wall to separate out breakfast. "I tune out the internal voices of my mother and my wife warning me not to take risks when I'm underwater," he explains.

Men migrating to Paracas, Peru, during the great El Niño–induced scallop boom times (Meltzoff et al. 2005) also physically drove themselves to collect more and more scallops, regardless of dive safety. Dying in numbers, their corpses were flopped up on the docks, draped in dark wet clothing. In the rush to return to the thick carpets of scallops, divers milling around these relative strangers paid about as much attention to them as to dead sea lions.

It's usually not a question of lack of training, but a pattern of suspending belief in invisible causes in science. I also witnessed death-defying dive practices among the Miskito lobster divers on Corn Island, Nicaragua (Meltzoff and Schull 1999). The Miskito continued diving despite having to mask the torture of painful bubbles lodged in their joints with heavy drugs or alcohol. Even those who had taken government courses in dive chemistry ignored the training. They attributed ultimate paralysis and death to traditional Afro-Christian spells worked against the victim rather than their own unsafe practices impelled by the lure of cash.

GALAPAGOS NATIONAL PARK INSPIRES ENVIRONMENTAL DEBATES

Villamil perches on old lava flows buckling and collapsing off Isabela's southeast shore. The volcanic islands of Galapagos are about 620 miles off continental Ecuador. To map gazers, Isabela's six volcanoes form a giant iconographic seahorse, the equator neatly bisecting the head (figure 2.1). Lava coats much of the terrain, streaming and cascading into the sea and under the shallows in frozen formations now decorated in soft corals and other rich sea life, home to sea lions, penguins, and marine iguanas. Stories circulate from the old days when men climbed up into the lava fields, never able to make it back to shore, their single pair of shoes and their feet shredded to bits. Even dogs need duct-tape booties to navigate sharp flows.

In a brief timeline of the Galapagos Islands (synonymous with the Park), Ecuador declared the whole archipelago to be the Galapagos National Park in 1959, one hundred years after Charles Darwin published *On the Origin of Species*. Galapagos became Ecuador's first national park, limiting human habitation to only 3 percent of the land of this Ecuadorean province. Direc-

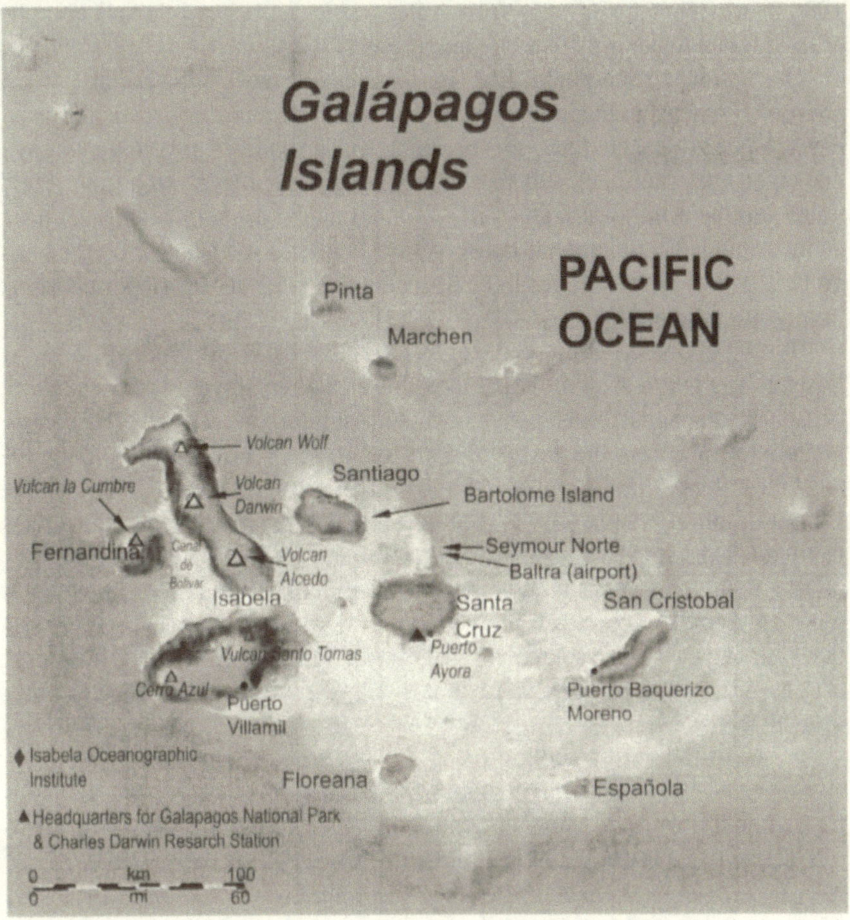

Figure 2.1. Galapagos National Park map. Adapted from base map from Wikimedia Commons (Eric Gaba).

tors were appointed by the national government. Rangers and officials began work in 1971, just as large-scale yacht-based tourism in Galapagos was getting under way.

Park conservation efforts included delineating visitor sites and tourist regulations on the one hand, and on the other hand starting to instill a conservation ethic among the few thousand residents. The Park asserted itself among the people by confiscating Galapagos tortoise pets and making it decidedly illegal to eat the tortoises.

In 1978, the United Nations Educational, Scientific and Cultural Organization (UNESCO) declared Galapagos a Heritage for Humanity. With the onset of boom-time export fisheries, the Park added on the Marine Reserve in 1986 that extends jurisdiction forty miles out to sea beyond the islands. The Marine Reserve controls fisheries and protects sharks.

The national government had the Park create the 1998 Special Law to address runaway population growth and its ecosystem crisis. In terms of enforcing the Special Law, the national and provincial governments have displayed little political will to staunch continuing illegal migration. Hard times and the need for a safety valve for continental economic upheaval have countermanded international conservation concerns over out-of-control migration. It has been a paper tiger Special Law, with its restrictions against immigration ignored.

By 2007, UNESCO put Galapagos on its list of World Heritage Sites in Danger. The national government took this as a direct threat to their economy because Galapagos tourism generates tourism for the rest of the country, visitors touring from the Andes to Amazonia. The World Heritage Committee voted it off this list in 2010, although the International Union for the Conservation of Nature (IUCN) thought it was premature because of threats from tourism, invasive species, and overfishing (BBC News 2010). The international conservation community continues to count on Ecuador's government to keep on track protecting the fragile ecosystems at stake. Thus the Galapagos, a key source of inspiration for Darwin and his thesis of origins and evolution, remains a hot site for environmental debates from tourism to fisheries.

In one of the most troubling scenarios, the human population—like invasive species—is escalating in numbers. During my decade of Galapagos fieldwork following the new millennium (2001–2012), the human residents of Galapagos doubled, with the population now hovering over 30,000.

Migration toward Galapagos's Apocryphal Wealth and Social Well-Being

Newcomers, as well as newborns, are quickly filling the 3 percent of available land in the Galapagos. Whoever comes sends for friends and family, and the cycle repeats itself. Births account for perhaps half the growth. Isabela on the western frontier still has the smallest population among the three main islands allowed to have human habitation.

With or without fisheries booms, people continue to enter Ecuador's economic promised land, often illegally. Galapagos has a 75 percent higher minimum wage, given the imported cost of living. Even doctors earn significantly more here. The vast majority have integrated into the two big towns of Puerto Ayora on Santa Cruz and Puerto Baquerizo Moreno on San Cristobal (see figure 2.1) where there's varied work, especially in construction, transportation, and other tourism-related jobs.

Employers tend to profit by using illegal migrants, avoiding contractual responsibilities. Many can wangle the legal temporary three-month residency. Then, upon expiration, they slip unnoticed into society. They only remain illegal until they can locate a marriage partner or a job contract. Both bring residency status.

Beyond economic salvation, newcomers end up treasuring social well-being summed up in clean air, peace, and security for an optimal family life. Miguel is one such migrant father who does construction. Raising three young children in Puerto Ayora, he sums up local sentiment saying, "We live in tranquility, so different from crime-ridden Guayaquil. Here, children are safe and can play freely. Air is pure. Despite higher prices on things shipped over from the continent, I can provide for my family."

Single women have tended to come to the Galapagos towns of Villamil, Puerto Ayora, or Puerto Baquerizo Moreno to visit relatives and help out with the children, cook, and clean. Most women gain residency by marrying to enter the mainstream. If educated, they end up filling more professional jobs. If not, they might work in shops and restaurants, bars, and even the clubs.

From the perspective of women who've married and made their home in Galapagos, another potential downside of the fishing booms has been men's habitual stop at boom-time brothels before heading home, pockets significantly emptier. The men, whether local or newcomers who come to get rich, have wanted to indulge in "continental" prostitutes—from mainland Ecuador—whom they have seen as the height of glamour. Besides, there were few other women to choose from. Some women of the night married Villamil fishermen and became integrated into the community; some saved their money to return and build nice houses back in Guayaquil.

"When a man married his brothel sweetie then returned to the brothel, we'd be razzing him to go home, since he already had his doll waiting for him," says Julian, remembering Villamil's macho days of power from money.

Villamil, with roughly 2,500 souls, is the only Galapagos community where life has been revolving around fishing. They also own the most fishing boats per capita, according to Villamil's fisheries co-op—one of four founded by archipelago fishermen. The Galapagos's first fisheries export market boom in lobster drew in migrants from the continent. But it was the tales of obscene wealth to be won for pepino diving in a second boom that lured in many more.

The heart of the pepino boom pulsed from 1993 to 2003. A Park photo of an unbelievable sea of landed pepinos (Galapagos National Park 2003) shows how in one decade, Galapagos fishermen could lift forty million individual pepinos out of the sea. Fishermen did not want to be managed by the

Park. Coordinating through their co-ops, they successfully blocked Park management through strikes that threatened the bookings of yacht tourism.

The Ecuadorean oligarchy tends to own tourist yachts in Galapagos as well as industrial fishing fleets in the continental port of Manta, with members of the same family often holding high national political office. Galapagos being on the radar of international news, it is easy for fishermen to make a huge news splash that can harm tourist bookings. Fear of this brought the oligarchy into alliance with the fishermen. Thus divers accomplished a political move against Park management by striking. Then the oligarchy, wishing to protect their vested interests in Galapagos tourism, used their national political positions to block Park fisheries management.

With a dearth of management, pepino was reaching a point of diminishing returns, and divers were taking greater risks to reach one more pepino always deeper down. Finally, pepino petered out in 2007 through overfishing. Divers could not land enough pepino to make it worthwhile for the buyers, so there was an economic crash. Only at this point could Park managers and conservationists step in and declare the pepino fisheries closed until pepino recuperated. Everyone expected several years without pepino income, and diving halted without a strike.

While the pepino boom was winding down, Isabela's municipal government and its mayor started positioning their community to gain a piece of the archipelago's tourism pie. The Park had been developing more tourist sites around Villamil, from boardwalks to beautifully signed trails trimmed in lava rocks. The municipality knew yacht-based industrial tourism was booming and predicted that the next phase would be land-based tourism that could include them.

The municipality diligently began work in earnest to make Villamil a desirable destination, from redesigning the central plaza to creating a large stylish wharf system with designated tourist landings. *Isabela creseporti*—"Isabela grows for you"—was engraved on golden-hued wood benches and placed around town in prominent locations. Isabela people, including the more prosperous fishing and farming families, developed various levels of hotels and restaurants and started local tour agencies in readiness for the newest type of boom-time.

Access to the Galapagos environmental pie is divvied up among three umbrella interest groups: fisheries, tourism, and conservation. Fishermen discuss this division and feel they deserve their share based on access to marine resources, having used strikes to ensure maximum catches of lobster or pepino.

The national elite who own Galapagos's prosperous yacht tourism allied with fishermen to block Park conservation efforts. Minimizing management and conservation has ended in hastening the very decline of lobster and pepino that leaves Isabeleños empty-handed, yearning for their own land-based tourism.

Transition from *Fincas* to Fisheries

Sundry people have been drawn to the relative isolation of Isabela. Since Isabela is the least-visited and least-developed island, Isabela people have stayed relatively insular and proud of their independence. Isabeleños first came to settle the highlands, farming the rich volcanic soil even though it was rainfall dependent, unlike the three other possible islands they could have settled.

During World War II, the United States made military surveillance bases to keep an eye on the Japanese and secure the Panama Canal. To counter the American desire to annex Galapagos after the war, Ecuador encouraged colonizationn of Galapagos. A few *fincas*—farms—were already up in Isabela's highlands. On the heels of the Americans, the national government established a prison colony, shipping over convicts along with prison guards and their families. The prison divided into four camps, with the least dangerous criminals in the highlands, laboring on the fincas.

Villamil served as the tiny beach port for the finca people plus the prison community. Where the U.S. military had erected a small radio operations base near the coast, the prison set up the sadistic top security camp where men, shirtless in the blazing sun, barefoot among the spiny plants, piled rocks to form a towering enclosure perpendicular to a natural escarpment. It was considered a natural prison, harsh and without water, nowhere to run. The Wall of Tears has become a Park tourism day trip outside Villamil. Now, neat lava steps lead up the escarpment near the point where the high trapezoidal wall abuts it. Repatriated giant tortoises graze in the scrub below, chewing on cacti around the World War II concrete foundations. I shiver standing under the wall's deep shadow, watching lava lizards hide in the crevices. The site feels similar to Pisagua's natural prison (Meltzoff et al. 2001), where emotions ricochet off the politically painful history of Chile under Pinochet.

The prison system shut down in 1959, the year the Galapagos was declared a national park. News of the cruelty on Isabela had spilled into the press after a spectacular prison break. In a role-reversal theatrical, prisoners acting as guards wielded actual weapons. During their escape, townsfolk took refuge in the Catholic Mission in the center of Villamil. Some managed to hijack boats back to the continent, only to be apprehended. But the release of stories of death-defying conditions aroused public opinion and brought an end to this shameful period.

Dating from the prison period, the Mission is the oldest edifice on Isabela. Its Doric columns, carved by captives out of lava blocks, stand in front, and one now bears a historic plaque commemorating the prison's termination. Several of the guards and prisoners stayed behind, and their descendants are *Isabeleños*—Isabela people.

Isabela could provide land and quality of life for those who arrived. In their own words, the epitome of social well-being for these continental colonists, many being from peasant poverty, was the boon of plentiful wild meat. Beside the Galápagos—the endemic giant tortoises—feral cattle, pigs, donkeys, and goats roamed in abundance, free for the taking. Diet no longer includes tasty giant tortoises that have become highly illegal/off-limits. However, some old-timers still eat it for supposed curative powers. The Park controls 97 percent of Galapagos land, and rangers are supposed to register hunters heading out into these uninhabited wilds. Park rangers say they know the rare culprits continuing to supply tortoise meat. But when a hunter returns to Villamil, it's not hard to conceal an occasional leg of *Galápagos*—slang for the giant tortoise—among butchered haunches of boar and bull. Only a handful of dedicated hunters are based in Villamil; it's a specialized passion and livelihood.

Descendants from most fincas were attracted to Playa—"Beach," the local nickname for Puerto Villamil—by the same fisheries booms that lured newcomers. New arrivals married into those families who had first settled the fincas. Continuing the famous Isabeleño frontier identity, they proudly have referred to themselves as renegades—outside the reach of government control—far more than the rest of the archipelago.

About 20 percent of Isabeleños still live in the highlands. They often come down on weekends or commute. Twice daily, a Villamil bus loops up and around, dropping people in the three settlements or at trailheads leading to secluded fincas. Isabeleños hold on to a pride in highlands land ownership, wrapped up in a sense of cowboy heritage that's on display during Villamil's annual Carnival events or political rallies.

LETTER: MEAT AND HORSEFLESH

Hey Kim,

Thinking of you and your Falkland cliff faces full of penguin chicks and gaming albatross. For me, the landscape shifted from Miami to marine archipelago. But unlike your Falklands, there's a weird mix of equatorial cold and warm water species in Galapagos. Penguins and albatross are included. Instead of your famous merino wool sheep, the fincas here focus on meat.

Checo, a local friend who lives between Isabela's joint worlds of fincas and fishing, takes me to visit the scattered farms with cattle and horses, and pigs and chickens running around scrabble yards. Hunting dogs bound at Checo's car to check out strangers, and are only willing to receive pats once the owners reassure them. Nothing as sweet as our pet Aussie shepherd, Mango!

Approaching the second largest crater in the world in the cloud-covered ecological zone, Checo says the next place belongs to one of the very old men who'd been a prisoner back in the brutal prison period after World War II. He'd never left, and his four grown children and their families live in town.

Checo asks the old man if he can buy a chicken. After a nod of approval, Checo chases one red hen, managing to corner her up against the wooden house behind a barrel. Swinging her by the neck, he pops her feathered mass into an old rice sack. A few dollars are exchanged. Dinner will be organic like almost all meat on Isabela.

Doña Anita Chapi at eighty years old still grows pineapples and sugar cane, and raises chickens on her finca up in the Santa Cruz highlands in the settlement of Santa Rosa (figure 2.2). I first met her a decade ago as the mother-in-law of my guide, Rene Sangolquin. We were camping out on her family land where giant tortoises come to copulate and roam in the fields among the cattle and lava rocks.

Doña Anita's husband, Segundo Chapi, had founded Santa Rosa on land given by the Ecuadorean government in an effort to colonize the Galapagos

Figure 2.2. Doña Anita Chapi, Santa Cruz pioneer, in her house with corn drying. Courtesy of the author.

after World War II to cement their national claim to the archipelago. Don Segundo recruited colonists on the continent and brought them over to this potential haven. But only a handful of the original families lasted in the rugged isolation. Don Segundo, himself, was often away on Santa Rosa business. Doña Anita recalls having to give birth to several of her children up on the finca all alone. She said she was unable to make the hard eighteen-hour trip by donkey down to Puerto Ayora.

Of the pioneer families who actively maintain their love affair with horse flesh, Modesto Tupiza, both father and son, resonate for me. They keep a number of horses grazing on their lands, and in Villamil's rising climate of tourism, the demand for horses to ride along the rim of the giant crater is growing. They have become one of two major operators for horse tours, and I think you would approve of their good care, despite being tour-giving trail horses.

After a couple years riding around the crater on strawberry Mora, their smooth Peruvian-gaited horse, I had asked if I could adopt her so she can retire on the finca, skipping tourist rides. Modesto junior said he loved Mora too much to agree to any kind of adoption, even though I only show up a couple of times a year. I guess my offer was too odd. I'd had a great fantasy going. Seeking out Mora last year, she wasn't on site, so Modesto senior offered me another of his great mounts, Hilo de Oro, with a similar sweetness and smooth speedy gate. But by this year, Mora had died, said Modesto junior. I'd caught up with him driving around town in his wooden open-sided tour bus, called a *chiva*—"goat." Inside, passengers sit on long wooden benches facing forward. On each of the outside panels, he has vivid hand-painted fire-spewing volcanoes, flamingoes, tortoises, and Isabela penguins.

Worthy of a classic Western, this morning a handsome cowboy and his sidekick gallop down main street, hooves kicking up the sand. Astride a high jet horse the color of his broad mustaches and hat, I feel like I'm in a flashback. Perhaps this hints at a flash-forward to a new era of Villamil tourism when horseback riders descend from the volcano to gallop along the broad beaches.

What an angle you got on that rock hopper penguin's feathered eyebrow! Was he nesting on the cliff behind your house on New Island? Missing you and sending wishes for your luminous life and wildlife photography.

XOXO Sarah

STORIES FROM COFFEE AND CATTLE TO THE COASTAL ENVIRONMENT

Galapagos coffee is recognized as organic and delicious due to the lava-rich soil. For decades, coffee has been a Galapagos cash crop. The father and son of one Isabela family dominated initial coffee marketing and profits. After they died locals grew for themselves, and Jose Gomez became head of the Isabela coffee growers. Jose's own unroasted beans won top awards at a recent competition for all the Americas.

Jose is lithe and wiry, with a full shock of gray hair. He is a highly respected character who came as a policeman. I head up to the settlement of Esperanza to visit Jose's finca and see the red coffee fruit ripening on the bushes—and of course drink the dark extract. It's brewed by soaking fresh grounds overnight in cool water, which eliminates any acid and concentrates the coffee into an infusion easy for carrying on boats or camping trips. Jose's stirring his dried pale beans over a wood fire in a pan balanced on a grill (figure 2.3). Hand-roasting is the time-honored technique, yet Jose is purchasing his first mechanical roaster, due to arrive soon. A grin lights his face as he launches into the story of how life brought him to this finca.

"I arrived in Isabela in 1973 as a young policeman. There wasn't crime, just chicken theft. I fell in love and married. My wife's family already owned this coffee finca as well as house land in Playa. You've seen our place in town near our son, Byron. He came home after university and is helping me market our coffee. His first job back here was managing the fisheries co-op, and then he won the election for vice mayor. Pedro, our other son, is the nature man and one of the few young men to become a full-time hunter for Villamil. Wild meat is cheaper than finca-raised meat, and Pedro has regular customers who prefer the free-ranging life and quality of the meat.

"In my early days, everyone enjoyed bringing down free meat, besides the cash boon for catching live cattle and goats for the Guayaquil market," said Jose. "The government technically owned all those feral animals, but people could easily catch them for good money. Police issued 'farm animal ownership' tags from the Galapagos Provincial Government on San Cristobal Island. These tags let you show that the animal belonged to your farm. However, wild cattle weren't roaming far afield then. Inside eight days, a local could catch a cow and herd it to his finca, bathe and tag it, and fatten it up to sell as his own domestic cow bound for Guayaquil. Everyone knew this was happening and nobody cared what animals originated in the bush! A couple of regular buyers came on Guayaquil's monthly cargo ship that could carry up to 80 head of cattle and maybe 300 goats to market. One cow earned a man the equivalent of almost a week's wages as a day laborer. A goat brought in as much as a couple days' wages, and those goats were just wandering around the yards in Playa for the taking. What a temptation! It was part of the local way of life and considered a big benefit for living off the beaten track."

We're taking turns scratching the ears of a floppy house-pet dog and a hunting hound who's recuperating from a goring accident. Jose says, "I loved going out hunting like Pedro. Pedro works our pack of dogs to track through

Figure 2.3. Jose Gomez fire-roasting his award-winning coffee beans. Courtesy of the author.

the dense scrubbed backcountry between Isabela's two southern volcanoes, staying off the sharp, barren lava fields. He and one or two companions ride horses and mules, going fast and deep into bush terrain riddled with collapsed lava holes. They sleep in permanent tent camps they've set up for shelter and must keep stashes of bottled water since there're no natural water sources up there. It's dangerous. When a startled boar or bull charges out of the scrub, they have to scramble up the nearest tree. Sometimes Pedro carries a rifle, other times he just follows on the heels of the dogs, hunting with his machete and lasso."

Returning from one of these three- to four-day trips, I see Pedro has brought down a couple of young bulls and a couple of wild pigs, partially butchered and slung like saddlebags across his mules. Pedro had made room for more meat by riding high astride a carcass on top of his saddle. Hunting is a marker of manhood and provides a deep sense of satisfaction and well-being, like fishing, say the hunters and fishermen.

"Early hunters on Isabela had easy pickings close by, not such adventure as we have!" says Pedro among his companions, as they happily compare themselves to the old-timers.

Gregorio Caiser is one of the old-timers. He reached Isabela by continental cargo boat, riding in the hold with others seeking their own land and a way out of poverty. Now over ninety years old, Gregorio, bent over but strong, still makes daily rounds of his finca wielding his machete. Oranges are scattered everywhere in the shade of his gnarled old citrus trees that still drop plentiful fruit from great heights.

As he peels fallen oranges with his machete, passing juicy pieces to me, Gregorio says, "Raised on a farm, orphaned, I had nothing as a teenager. So I followed the word of a relation to try Galapagos where I could earn land. Maybe it was around the end of World War II, when American soldiers were pulling out. I worked hard, clearing my own fertile soil, planting yucca, coffee, and orchards of these oranges and avocados."

"When Carmen came to Isabela, I swept her up!" says Gregorio, proud of his mate still by his side. "She was fleeing an unhappy marriage in Quito." Carmen—a city woman—was following the familiar course for women willing to brave the unknown. She'd entered a rugged world, and with Gregorio, they began rearing their family in the highlands (figure 2.4).

Gregorio's is considered by locals to be a solid family. "I wasn't into cock fighting, gambling like most men. I preferred keeping lots of chickens, and we loved raising pigs—my specialty. Among our five children, only one is dedicated to finca life like me. The others live in Playa. Carmen and I go down on weekends to stay with Manuco, who became a fisherman, Lidia next door with a grocery store, and the rest," he says. "So we'll see each other again at Manuco's on Saturday."

Figure 2.4. Caisers with their cat living on their Isabela finca. Courtesy of the author.

Manuco Caiser epitomizes his generation's livelihood transition as Isabela shifted away from the finca into fisheries, and now tourism. Manuco is in shape as an excellent diver and has an engaging personality, warm and high-spirited. He has learned to survive in the highlands environment as well as the marine ecosystem. Born and raised by Carmen and Gregorio on the finca when Isabela life still gyrated around the finca, Manuco describes his childhood in the early 1970s.

"Just four or five families were living down in Playa. The rest of us—about fifty families—were mostly divided among three highland settlements. But when the cargo ship dropped anchor every month or two, the finca families like us would converge on Playa for a few days of socializing and trade. We'd shop in the few stores for whatever goods were being offloaded, especially dry goods and staples like sugar, rice, flour, cloth, tools, and a few treats.

"I loved it when all the kids played in the surf, and some of us boys caught lobsters hiding in shallower crevices. We'd sell them to Playa people for spending money before returning to the fincas and chores like planting yucca, weeding, and looking after my parents' pigs. And it was great going hunting on horseback, taking our dogs."

Like many young men, Manuco moved to Playa during the first fisheries boom in lobster. He is a perfect example of how the boom triggered Isabela's shift from the finca to harvest the sea from Puerto Villamil. Manuco, soon a consummate lobster diver, met and married Mireya Rogel, who had become one of the first lobster buyers for Guayaquil's export market.

Mireya is a dark beauty whose vivid eyes flash humor and intelligence. In her, Manuco found an extraordinary family and business partner. As Isabela has been changing inexorably from fisheries into its own land-based tourism, Mireya and Manuco have been slowly and literally building up Casa Delfin, their own family hotel, on land from his father, Gregorio. Mireya likes to observe and paint the nature of Galapagos, and her work decorates Casa Delfin.

Mireya finally stopped exporting lobster to Guayaquil after the fisheries petered out. She only runs a large freezer for seafood sales out of their hotel kitchen. Over a treat of lobster tails at Casa Delfin, she and Manuco describe hopes for their tourism, and how the Park has been hiring Manuco in a new role as an expert senior fishermen to test fisheries options, such as lobster traps, that might supplement income. This is a reversal from boom-time dependence on fisheries.

In another flip, fincas can now offer a diversification from fishing and tourism. For those like Manuco, who always maintained a finca—hiring minimal labor while fishing—they can now draw on cattle for capital to develop tourism. With the price of raised beef good in Galapagos (it's illegal to export meat) cattle can offer an economic safety net. Manuco has sold some of his cattle to cover hotel loans used to add on to Casa Delfin's top floors and balconies with ocean views.

Manuco points out that even some finca folk are joining the trend into tourism. He says, "My brother, Carlos, who always preferred finca life over Playa, is supplying the coastal clientele of restaurants and families with produce from his commercial vegetable and fruit garden. His farm is right on the road as it begins to enter the highlands. It's a good location, not too far from town. And Carlos and his wife are constructing an open-air restaurant with a weekend grill that is already attracting people from Villamil."

Lobster Boom as Seen through the Eyes of Mireya

Like many Puerto Villamil women, Mireya came here single from the continent to visit and help care for the kids of a relative—the sister of her brother-in-law. "I joined Villamil when there were no more than 250 people. It was 1985, on the verge of the export market boom in lobster. I was an independent seventeen-year-old fulfilling family duty to help out Flor, when I boarded that cargo boat to Isabela," Mireya says.

Newly widowed by a dive accident, Flor had found herself alone and quite pregnant, with two young kids, and forced to accept that her own husband was among that year's divers dying for the dollar.

Mireya continues, "The young couple had arrived in Villamil a couple of years earlier to fish, joining a number of Loja Province Ecuadoreans who had migrated here, one neighbor telling another. Flor's husband found a job crewing for Oscar, a Guayaquil fisherman running a mother boat out of Villamil. The mother tows a little fleet of open boats—*pangas*—used by the divers. Oscar pioneered harvesting lobster in Isabela with his mother boat, and bought up the catch of others. He learned firsthand the location of good grounds and befriended the locals who dived them. Oscar accurately predicted that lobster landings would soar to boom proportions over the next few years, and wanted to position himself accordingly."

Feeling sorry for Flor and understanding the need of his young diver's widow, Mireya says, "Oscar recognized a mutual opportunity. He set Flor up as his Villamil lobster agent, and headed back to Guayaquil to refocus his own energy on building a lobster-processing export business."

Flor swept Mireya up into her new lobster venture. "Luckily," says Mireya, "I'd had lots of market experience. After an earthquake shook my family out of Loja, I had grown up in a commercial town on the Peruvian border. When my father abandoned us, I started working in the open-air markets and boutiques even though I was only twelve.

"Starting out at dawn, Flor and I would find guys to roll our ten seventy-gallon barrels brimming with frozen lobster tails on ice, all the way down the sand road to Villamil's wharf. There, we had to hunt for a willing fishing captain who was crossing eastward to Santa Cruz and would agree to carry the lobster tails and Flor. Flor and her barrels then had to catch the once-daily flight from Baltra Airport to Guayaquil. She'd deliver the precious tails in person to Oscar's Guayaquil plant."

After a couple months of an increasingly difficult pregnancy, Flor decided to stay put on the Guayaquil side. She wanted to take advantage of a real hospital and ensure a safe delivery. Mireya says, "I was left with the reins of the Villamil lobster business as well as caring for Flor's two kids.

"Mother boats were pulling into Villamil after a dozen days at sea. Most mother boat captains ran a dozen-crew hookah diving from four open boats, each with three men: two divers and the one who ran the compressor and followed the bubbles," says Mireya. "A half a year flies by for me in Villamil with Flor still in Guayaquil. Our lobster business flourishes."

Besides lobster, love blooms for Mireya. "Manuco was one of the young divers helping me roll the barrels to the port, and selling me his catch. When he proposes, I decide to stay in this place. I loved it's free-rein attitude, the town's feeling of independence from authority so far away, the beach in front of our house with marine iguanas on the rocks, the salt air, lots of seafood and meat to eat, as well as my own market ability to earn great cash." Their children were destined to become Isabeleños, as Mireya adopted this identity, like the others who marry into the community. Mireya looks around Casa Delfin's ample family-style hotel dining room with the computer humming on the side table, and freezer droning in the kitchen, mentally comparing the changes in material lifestyles.

Describing material aspects of Villamil life when she first came, Mireya says, "We used a generator to keep the lobster frozen in six small freezers. Electricity was erratic and generators were mandatory to store perishables. Energy could and did cut out at dusk. People strummed guitars instead of turning on the radio or TV. We had only one bakery that would open and sell out its bit of bread at 4 PM, so we'd have to wait in line. Most of us rode bikes to get around town because there were few motor vehicles. It took us five hours each way to reach the big sticks of Puerto Ayora on Santa Cruz. We'd have to hitch a ride with fishermen, bouncing in the bottom of their open skiffs—*fibras*—which weren't equipped with seats and awnings, let alone plastic sheets to shield us from waves or rain, for another fifteen years. Comfort didn't come until the millennium and the start of local inter-island transport business.

"People began streaming into Villamil for the lobster diving, and our population had doubled by the lobster peak in the early 1990s. All during the lobster boom decade, 1985–1995, I continued buying lobster. A small plane had begun service between Villamil airstrip and Baltra, sometimes making three flights a day loaded with lobster to connect with the big planes to Guayaquil. This made it a little easier at this end.

"Flor eventually returned to Guayaquil to educate her children, and only came back much later to make her mini-market 'Lizette' and live in Puerto Ayora, which is a more urban lifestyle," says Mireya.

Sending children back to the continent for high school or university is the common reverse-migration pattern for families who put emphasis on education and can afford it. Mireya and Manuco spent their earnings to send their two older children to the continent for university, and have plans for the youngest.

Mireya says, "I kept going as a buyer, keeping three freezers at home to be ready, and staying on call at any hour, from 6 AM to 10 PM, hard work. I bowed out of the lobster buying business in 2006, as quantities and signs pointed to far fewer lobster but hope for our local tourism. Manuco and I keep using the three freezers to sell his personally caught lobster and fish to tourists and neighbors.

"Even into the new millennium, refrigeration in Villamil couldn't keep perishables without running a home generator. Now because of the current constant supply of electricity, lots of families are investing in freezers," says Mireya. Having a home freezer these days is considered an upscale, quality-of-life investment for storing their beloved meat and fish. As Manuco puts his cleaned catch into the freezer from the morning's spearfishing off the beach, Mireya pulls out some frozen fish for our lunch. She's also boiling soybeans with sugar to make delicious sweet milk.

Loan to Fishermen a Loaded Gun

Mireya and Manuco continue their personal experience of fisheries booms. "In retrospect," says Mireya, "Villamil didn't really know what crazy was until the decade of the pepino boom, 1993 to 2003. We ended up with ten times the people from when I first arrived. Men, women, and children joined in the pepino boom in Villamil from diving to processing. The least paid member of a pepino crew, the one who stays up in the boat to run the hookah compressor, could earn $100 a day, a fabulous income," says Manuco.

"Even eighth-grade boys could work and bring paper bags bursting with bills to school, inviting their teachers out to lunch," added Mireya.

In a typical initial flash of short-term pleasure-spending, locals lavished much of their first-year pepino wealth on fiestas and clothes, or the proverbial act of lighting cigars with $100 bills. Nobody could believe it would last.

"At first, social life revolved around spending these bizarre amounts of cash in once-in-a-lifetime flings. 'Squandered' was the word used by conservationists against us," says Manuco.

Four women set up bars, bringing in continental prostitutes, considered the height of sexy and prestigious by local men in those days. Divers made these establishments their pit stops with plastic bags bulging with cash, before going home. A number of the women married into the community and faced the same fate as the other wives as familiarity wore off their exotic splendor. But some of the boom-time prostitutes carefully invested their earnings into large Guayaquil homes, where they reside.

Over subsequent years, as the money kept rolling in, families started to trust the income, and even to expect it. Wives of fishermen, in particular, began insisting on family investments, wanting houses and perhaps a little shop in lieu of their husbands' yen for liquor, prostitutes, and grown-up toys and engines that degrade quickly in the sandy salt air. I watched the first act of a woman provincial governor of Galapagos, which was to raid brothels and deport disreputable women, pleasing her female and church constituencies no end.

Isabeleños developed a more consumer-oriented lifestyle, complete with large-screen videos. Sustained years of cash elevated everyone's expectations, and locals embarked on feeling the need for "cargo"—my favorite term for industrial goods from the global economy.

Villamil's level of family alcoholism and violence, and lack of parents educating their children, is a community concern. Those involved in education on the island point to boom-time marriages between men fishing and ladies of the night as having contributed to this situation. There are parents who cannot afford to send children away to better schools on the continent, or conceive of it, let alone help with homework. There is no reverse immigration in these families. Life will be carried out on island, for better or for worse.

Villamil fishermen had formed one of the four Galapagos fisheries cooperatives, Nuevos Horizontes de Isabela, which reported a tripling in the number of boats between 2001 and 2006. Ironically, it was the government that spurred this surge by lending fishermen money through the National Development Bank. In 2001, still in the midst of the lucrative pepino boom, almost all Isabela fishing families were tempted by the Bank offers to take out loans. The Bank had come to them touting fleet expansion. It seemed logical to the hopeful fishing families, given the pervasive concept that growth equals good business practice. These were times of high-income catches, with the experience of two rewarding booms back to back.

When the Bank opened a line of credit for fishermen at the going rate of 18 percent, fishermen, who normally have scant access to credit, were excited to sign on. Wives had to cosign their houses as collateral, locking the family together into debt. Couples felt this was their chance to be savvy and invest in capitalist expansion to secure more pepino profit.

Fishermen built more boats with the loans, increasing their size within legal artisanal limits to still qualify for fishing rights inside the Marine Reserve. However, building larger artisanal boats unintentionally enabled them to illegally market forty miles out to sea, on the border of the Marine Reserve. Herein lay temptations to smuggle pepino and illegal shark fins to the Manta-based industrial fishing fleet owned by the oligarchy, as well as to foreign-owned fleets trafficking to Asia.

A spate of poor pepino harvests in the later years of the boom, as well as scant secondary catches of lobster, followed on the heels of this gamble. Fishing families worried. The small fishing community sank into a sense of economic despair. Even three years later, half the families were under the gun, unable to pay off the loans. Reality had run contrary to their high hopes that capitalist expansion of fisheries—increasing "effort" in fisheries parlance—would yield more wealth.

Shouldered during high times, these debts goaded the fishermen and encouraged them to strike against management, who they felt would limit their ability to save their homes. It produced pressure on the men to dive to ever-greater depths to collect the thinning pepino, despite the risk of death.

Wives of Fishermen Unite

Toward the end of the pepino boom, conservation interests were pushing fishing families to find alternatives, including ways for wives to earn income. Mireya is a leader in these efforts. She says, "The Park managers and their international conservationists wanted men to stop diving, and wanted fishermen's wives to form women's organizations to earn cash. In 2002, eighteen of us banded into OMAI—Women's Organization for Artisanal Initiatives. We sold food on the weekends, and by the second year, the Park and municipality gave us training in silk screening and painting. The Japanese government chipped in too, and we bought our big, hand silk-screening machine. We also started sewing stuffed animal souvenirs, thanks to another workshop."

OMAI women are silk screening their own T-shirt designs, tie-dying and tailoring sarongs and beach wear, and bringing to life stuffed animals of the iconic Galapagos penguins, marine iguanas, blue-footed boobies, hammerhead sharks, giant tortoises, and flamingoes.

"Four of us still really run OMAI," says Mireya sitting with three of her compatriots, sewing hot-pink felt into the stuffed flamingoes offered in four sizes, the smallest with pipe cleaner legs. We're sitting inside OMAI's workshop/shop, strategically located on the town's main coastal road. It's attached to the side of Casa Delfin, a fine location for walk-in tourists. OMAI also places their crafts for sale at Isabela's Park Tortoise Rearing Station store where tourists are bound to visit.

The other women's co-op founded at the same time has not been as fortunate. Pescado Azul—Blue Fish—was also initiated as an alternative income source for fishermen's wives to wean the community off pepino. Not surprisingly, Pescado Azul's leader, Emma Gil, was also an experienced business manager like Mireya, having run the fisheries cooperative before establishing Pescado Azul. For several years, Pescado Azul smoke-cured tuna, buying the catch of the minority of captains in Villamil's fisheries co-op dedicated to finfish fishing: ten compared to seventy who dived for lobster, octopus, fish, and the prized pepino. Finfish fishing remained a modest pursuit and the Pescado Azul supported them.

With Wild Aid (an NGO based in Puerto Ayora) advising them, they had success in marketing their tasty product through Puerto Ayora's upscale supermarket at the central wharf, and to the yacht tours to serve as a local delicacy.

Sadly, Pescado Azul stalled dead in the water with prize money from the prestigious 2006 Equatorial Prize. Around the equatorial belt, only five projects win annually "that raise the profile of grassroots efforts to reduce poverty through the conservation and sustainable use of biodiversity" (www.equatorinitiative.org).

Pescado Azul's proposed construction site for an improved processing plant within the community was deemed inappropriate by the municipality. At first, the Park promised to rescue Pescado Azul with a bit of land outside the main residential neighborhoods where the potential smell and noise from fish processing would be acceptable. It would have been next to the fisheries co-op's *acopio*—fish freezer plant—another promising project. Nothing came through, and both projects collapsed.

Isabela women have wished to alleviate financial pressure on their husbands, not only so they'd be less inclined to dive dangerously for scarcer and scarcer pepinos, but also to stall the practice of the ever-present, last cash resort of illicit shark-finning that can get their men in deep trouble. Galapagos shark-finning isn't legal, but it is lucrative, over time devastating hammerheads and other thick-cartilaged beauties. Galapagos fins only have to be smuggled to the continent, where the shark fin market is legal. Once there, nobody can tell if the origin is Galapagos. Ecuadorean law cannot protect the plight of Galapagos sharks since it allows the Asian market to buy up this luxury commodity on the continent. Ecuadorean sharks are considered a by-catch—not legally fished as the target species, but caught incidentally alongside the regular catch.

I've asked national fisheries officials in the main fisheries port of Manta about this shark-finning loophole. Their rejoinder is a basic tragedy of the commons—if we don't do it others will. Without a ban along the whole Latino Pacific coast covering multiple nations, it is pointless. A pan-Ecuadorean ban would leave neighbor states lustily catching shared shark stocks, so Ecuador feels justified in allowing trade in shark fins as long as their neighbors do. Therefore, smuggling the fins out of Galapagos is a continual temptation for Galapagos fishermen.

Unable to control fishing practices, including finning, international conservation programs have focused on building co-ops among fishermen's wives for alternative income. They lend financial and technical training. Globally, such microloans for women have had a high success rate in women's eyes, even if lenders want the capital to create bigger and bigger businesses. Staying small-scale, women can build critical emergency savings to maintain their families.

LETTER: RARE PINK IGUANAS AT OMAI

Dear Katie and Ed,

Imagining you with your rain barrel invention and vegetable gardening in Minnesota.

I'm back from a sunset bike ride to Playa de Amor—Love Beach. Large marine iguanas in red hues for mating are resting all over the rocks, males guarding their leks—territories—with vigorous head-bobbing and occasional sallies at each other.

Flamingoes are feeding in the salt flats behind Villamil. In the glooming, I glide up to Casa Delfin to visit Mireya and Manuco. The couple is alone in the storefront of OMAI, among the stuffed animals and silk screening workshop of the fishermen's wives.

Pinking shears in hand, Mireya is slicing hot-pink felt into long spiny ridges. I look at it quizzically, and she says, "This is not some flamingo. It's the world's first pink iguana. They've really been discovered on Isabela, so we're making them as our latest model stuffed animal."

Excited, I ask where I could see a breathing pink iguana. Manuco points to the wall map behind him, and says, "They only live at the tip of Isabela's seahorse nose by Volcan Ecuador, but that area is off-limits even to tour yachts. Only scientists can visit."

Many will yearn to see the pink phenomena in person. Yet, it's uncertain whether the Park should or would ever expand beyond the fifty-one official visitor sites that have been in place for decades. Perhaps it is better to keep pink iguanas in the imagination.

Warm hugs, Sarah

STRIKES CONTRADICT THE TOURIST RHETORIC

On the heels of a decade-long pepino boom comes not only the move toward women's co-ops, but, a stricter climate for pepino management. Regulations tighten by mid-2004, causing fears and a community-wide catch-22. Because their homes are being held as debt collateral, how can they afford to let management reign? Talk in the community revolves around ways to deter pepino management. A strike is in the air. Meanwhile, divers from Isabela and the other islands practice passive resistance by ignoring regulations. They continue camping and harvesting around the environmentally sensitive pepino nursery on Isabela's west coast—Canal de Bolívar. "Bolívar must stay open!" cries a fisherman. It's an unconscious rally for the freedom to dive Isabela waters. Latinos, who are frequently far more politically aware than people in the United States, recognize the double meaning, given their leader Simón Bolívar during independence struggles.

Despite dwindling quantities of pepino, especially critical in Bolívar's fragile nursery habitat, the Galapagos fisheries co-ops unite to strike against management in all three towns. The leaders coordinate via the improved phone technology. They demand the right to keep the nursery open and not be subjected to longer closed seasons and lower quotas. To prevent repetition of previous tortoise-nappings or office-trashing, the Puerto Ayora police unfurl barbed wire and barricade the sole road winding out to Park headquarters and the Darwin Research Station.

In Villamil, unhindered by police, fishermen build their own roadblock of old tires across the main road as it leads out of the central plaza. This cuts off traffic to the Park and station offices. Over the course of several days, fishermen play soccer on the sandy road by the tires, relieving the tension from otherwise standing huddled in conversation. They permit a pickup to pass through, honking loudly, loaded with their wives waving hand-painted banners and shouting support.

In solidarity with their men, the wives spend hours up at the house of another OMAI leader, preparing meat for a "Support the Fishermen" all-male barbeque that evening. They deftly slice hunks of freshly butchered beef into zigzagging ribbons. Wielding a knife too, I listen to the women's banter ranging to fears over how the Bank could confiscate their homes.

This sense of economic desperation remains an unspoken underpinning in public forums over management. Strikers underscore the contradictions of global development on Darwin's idyllic shores. Fishermen demonstrations inevitably draw international media that in turn harms yacht-tour bookings. Images of shouting fishermen waving machetes with hostage tortoises jar international notions of Galapagos nature unsullied by humans. This is the historic image of Darwin's islands, and it is the one marketed to tourists.

The rhetoric of tourist advertising purposely evades mention of the human presence, let alone the invasive species, such as red ants and rats, goats and cats. Galapagos is touted as an Eden where animals have never learned fear of humans. It's a place epitomized by visitors swimming alongside harmless sharks and sitting next to ground-nesting blue-footed boobies engrossed in high-stepping dance rituals.

Notwithstanding this rhetoric, precedents abound for Galapagos fishermen striking against attempted lobster and pepino management. Perhaps exaggerating the details, the world press adores juicy news of Galapagos tortoises under siege. Famous media reports include fishermen in Puerto Ayora in 1995 threatening to kidnap iconic Lonesome George, the giant tortoise who is the last of his subspecies. Again in 1996, fishermen are described as killing six Galapagos tortoises on Isabela: "in a successful bid to win higher quotas for [pepinos] then selling for a record $2 per critter" (*Miami Herald* 28 January 2001).

Strikes have been the fishermen's de facto form of lobbying national elite owners of the tourist industry to block marine management efforts. The reason is clear. The Ecuadorean business elite—whose families are synonymous or intertwined with top government officials—have been able to persuade fellow elites to back off biological proposals for stiffer fisheries regulations. The elite want to curtail strikes, which can cause negative press that hurts bookings.

During the ten-year pepino boom, divers legally land forty million individual pepinos, meeting the quota of four million pepinos each year. Despite biologists' annual pleas for stricter management, fishermen win lenient quotas and rights to the Bolívar nursery, and ignore "paper tiger" regulations—those without the teeth of enforcement. Their last strike before the economic crash of pepino is successful in the sense of limiting environmental control.

Leaders of the fisheries co-ops are able to sidestep most management efforts through their alliance with tourism interests. Refusing any nonfisheries alternatives, leaders demand alternative fisheries if they cut back on pepino diving. The Park, despite trials seeking alternative fisheries, can never come up with a different fishing practice that isn't environmentally harmful—like longlining—or another high-priced marine product in sufficient quantity to fuel a third Galapagos fisheries boom.

Divers continue driving down pepino numbers to the point where they cannot locate enough pepino at viable depths to fill quotas, and many are already diving dangerously. By 2007, buyers withdraw, finding the quantities of pepino not worth their while. With the economic crash of pepino, concerned international conservation groups are glad to see the Park finally able to impose a global ban. Because there are not enough pepino landings to bring in buyers, the fishermen accept reality: management is no longer the culprit and Asian export fisheries can no longer sustain the community. Men are no longer tempted into death-defying dive situations. Illegal shark finning—fins smuggled dry to the continent for sale to Asia—always an undercover option, may also be down. The ongoing pepino closure will stay in effect indefinitely until the Park determines through counts that pepino numbers have sufficiently rebounded. Without buyers tempting fishermen, fishermen leave pepino alone. This is predictable because markets make fisheries.

With dreadful timing, despite the end of pepino, the Bank renews pressure on the one-third of Villamil families still mired in fisheries debt. The

Bank threatens to claim houses used as collateral or lands (which the Bank doesn't want) or else prison (which also brings in nothing). This motivates some families to scrounge up cash and pay in dribs and drabs.

Sea Shepherd Story

During the high emotions of the mid-2004 fishermen's strike, six young *Sea Shepherd* crew members, incognito except for their prominent whale tale pendants and pirate punk style, are visiting Villamil for a few days while their anti-whaling conservationist ship is moored in Puerto Ayora. Galapagos is one of their global bases, and they keep an office onshore in this symbolic mecca for environmental protection.

We'd met when I'd brought my University of Miami students on board the *Sea Shepherd* in Puerto Ayora harbor, and toured their vessel with its massive captured fishing nets mounded on the foredeck, UN flag and Dalai Lama peace souvenir in the pilot house, and whale murals in the mess where they served us milk and cookies.

Our guide leans on their famous weapon pointing over the gunnels, explaining, "We fire chocolate and lemon-flavored pie filing from these cannon. And we go up against whalers or cut trawlers' nets, buzzing them with zodiacs," he says nodding to several of these rubber boats on deck. We agree it's a dangerous business, and one of the many forms of environmentalism. Theirs has a special niche and function in gaining popular imagination to alter the law.

A week later, out at the Villamil Tortoise Rearing facility, a couple of the crew half acknowledge my waves, stage whispering as we pass, "Pretend you don't know us. We don't want anyone to know we're here."

As the strike heats up, Isabela's fisheries co-op bans water-taxi travel between islands. The *Sea Shepherd* crew, who need to return to their vessel, cause a scene. On their webpage, they post how Galapagos fishermen are holding them hostage on Isabela. My class and I must also return to Puerto Ayora in a few days to make connecting flights. But I realize there would be diplomatic ways to get dispensation from the leaders to leave if we have to. *Sea Shepherd*, however, revels in the high drama, stereotyping thug fishermen. Being held "hostage" has cachet, serving their more extreme political agenda as international conservationists. I feel sorry for those in the community who have hosted the crew and are now being blamed by the co-op leaders for harboring troublemakers.

Penguins as Omen of Trade-Off from Fisheries to Tourism

Penguins perch on a particular set of Villamil harbor rocks, a short hop into the sea. Mysteriously in mid-2007, the same year pepino peter out, Villamil penguins disappear. This is not a good omen, as people turn toward tourism for hope. Everyone recognizes the endemic Galapagos penguins—*Spheniscus mendiculus*—as one of Isabela's tourist aces.

During a one-year gap, penguins remain a magnetic mascot for the fledgling local tourism. Penguins continue to swim across Villamil wall murals and hand-painted shop signs. OMAI women keep silk-screening penguins on bags and T-shirts for sale to visitors.

In this swing year between banning pepino and launching into local tourism, fishermen seem to be projecting a fear of becoming dependent on visitors instead of maintaining status as "independent" fishermen. Penguins had settled out of the blue in Villamil Harbor long before tourism, and in that lonely year without penguins, the men blame themselves and their own harbor tours for scaring away this precious bird.

"We all took tourists too close," rued Estefan, a young man offering tours in his handmade square boat rigged with a fanciful roof. "I'd edge right up alongside penguins preening on the rocks, so we could almost see eye to eye."

Villamil's folk explanation fails to resonate with experience from the most famous penguin-viewing place, Bartolome's iconic Pinnacle Rock. Penguins have never abandoned this popular spot with its daily tours. Perching on rocks often just an arm's length away, penguins don't act afraid of either the snorkelers in the water or the photographers clicking away from zodiac rubber boats lowered from the cruise yachts. They stay put on their favorite rocks whether molting or not (they do not swim while molting) and come over to someone in the water.

Penguins hunting for baitfish have used my body extended by fins as a large floating obstacle, herding the shoal toward me to separate one bite of quicksilver. Unlike sea lions, which try to provoke play, nipping at your fins or beaming in flat out straight at you, for a split second turn, penguins pop up in front of a person, unexpected, paddle on the surface for a breather, and soar off again underwater. On lucky days, I've had penguins simultaneously swimming by with sea lions, turtles, and white-tip sharks, showing off the uniquely Galapagos combination of creatures from disparate climates.

As suddenly as they'd vanished the year before, penguins resettle Villamil harbor. Their return coincides with a serious shift into local tourism. Villamil fishermen once again take visitors to watch the tiny colony, only now the men stay several meters away, respecting a new semicircle of buoys.

Fishermen are being forced to reevaluate ways to earn money. They dive lobster and spearfish, and do minor tourist-related work. Most fishermen, who maintain a variety of jobs, see offering harbor tours or renting a bed to visitors, as a realistic means of supplementing income.

Trade into Tourism

Eliminating fisheries has been one international conservation goal. Before the pepino crash, the Park angled to reduce the number of fishermen by proposing an exchange of fishing and fishing-boat licenses for the high-value yacht tourism licenses. The catch is that once signing out of fisheries, the door is forever closed to a man's future generations.

These yacht tourism licenses remain well beyond fishermen's reach. The capital investments are enormous. Without outside partners who intend to own and control the operation, it's a fantasy. Licenses haven't materialized on Isabela except for a handful of fishermen who have become what are called the "faces" for outsiders with deep enough pockets to jump through an elaborate and costly set of paper hoops, and then to still be able to invest in tourist-scale vessels. In reality, the trades have also been minimal because of the Park's stalling—a pattern found throughout this book of government causing frustration by not following up on promises that might empower fishermen.

Up until recently, tourists and fishermen have informally found each other, sharing the desire to go out together for the day. Isabela now has an official Park list of fishermen allowed to do this kind of tour under the rubric of *pesca vivencial*—one-day fishing trips that never touch land. The list is still largely ignored locally, as temptations are high, fueled by feelings of self-righteousness. Who doesn't wish to show their new friends and guests a cool fishing spot, one historically felt to be their turf, and perhaps earn extra cash for the effort?

Fishermen Co-op Denied Self-Management and Promised Tour Permits

"Fishermen know the whole island, but especially the western shores rich in pepino. During the pepino years, even after camping was outlawed by the Park, most continued to sleep onshore, boiling pepino to cure it," says Manuco. With the total ban, the campsites of fishermen are returning to nature, and in the future fishermen are supposed to sleep onboard their boats (figure 2.5).

"Without pepino, the fisheries co-op has been asking the Park to pay us to guard our fishing areas, but the Park does not trust us. They call us the fox guarding the henhouse," says Manuco. This is normal, nervous top-down management practice. Yet, co-op members could be motivated to guard their spots along this vast lava coast. The Park doesn't have the manpower.

"Coca Cola did hire us via the co-op to perform major cleanup circling Isabela," says Manuco. "In late 2010, almost one hundred fishermen, three to a boat, divided into groups and swept the shores. We even clean underwater at the three official Park tour yacht sites on the west coast of Isabela. We

Figure 2.5. Pepino found by Manuco. Courtesy of the author.

scoop up tons of tourist garbage, anchors, and tin cans, glass, and plastic bottles. And everyone comes out to clean up Puerto Villamil. We're happy earning over $800 each for about two weeks work."

Legitimacy in the eyes of the Park will be the big issue when the pepino fishery reopens. In a small town, who dares kick out the older men or the boat owners who no longer fish? When everyone wishes to jump back into the water for the precious pepinos, the key will be who is up-to-date in cooperative dues and legal permits to fish.

The one-third to one-half of co-op members who had kept diving were beginning to see pepino repeppering their underwater terrain in 2011. At a Park meeting with the fishermen, the women came out en masse to petition the Park to reopen pepino. One third of their houses are still in jeopardy, collateral for pending Bank loans to fishermen.

LETTER

Dear Veronique,

Again, it will be wonderful to work with you as my teaching assistant in Galapagos. I'm in the archipelago and it's that time of day. Golden rays slant across Puerto Ayora's deep protective harbor where myriad yachts moor, from masted beauties to rusty regular supply ships to those twenty-meter motor yachts come to switch groups of tourists to a British royal vessel.

Almost 4 PM and it's time to check out what today's fishermen are landing and selling at their small wharf—Pelican Harbor. Pelican Harbor is the former site of the main freshwater well that the few people who lived here over the first sixty years of the twentieth century used for their water supply. With more people, the well went dry, but to add tourist color, a mock wind generator and water pump have been erected by the spot, joining the company of an old giant green plaster sculpture of the iconographic giant tortoise.

Strolling along the bayside road, I pause to take pictures of a Darwin finch perched on an "I Love Boobies" T-shirt hung in front of one the endless tourist shops. Veterinarians visiting from the United States recently issued leashes, so dogs are out for walks instead of bounding along the rocky shore chasing the marine iguanas trying to bask, and worrying them to death.

Six children are clambering around fishing boats moored at Pelican Harbor. Drawn by their laughter, I head out along yet another new piece of tourist infrastructure, a neighboring boardwalk leading from the shore road cutting out through fringe mangrove and over the shallows. Several parents are leaning on the railings in conversation, occasionally one calling down to play fair.

Diego, a father of two of these kids, holds up his unfamiliar camera. Pointing to the action below, he asks, "How can I take a picture?" With the light lengthening, I explain basic aperture/time open ratios and show him settings. Diego explains, "We moved to Puerto Ayora four years ago from the Andes because I could do construction work, and it's safe here, much better than at home."

Soon, everyone wants a family photograph, the kids jostling up the steps to the boardwalk to bundle around their aunties and parents. "Say Whiskey!"

From a closer vantage point at Pelican Harbor, I watch Jose Luis, a muscular fisherman in a striped shirt, with short black hair and a broad smile. He's gutting his catch with his animal audience in attendance. A great white heron poses on the cleaning table, and several brown pelicans perch up on the zinc sunroof, ready to swoop down if the flotilla of juvenile pelican and ancient brethren on the ground miss a morsel. One young sea lion has learned to hang out nearby and hops up the wet cement steps, coming right over to Jose Luis to nibble a morsel of guts from his hand. "Here you go, Pancha, what a good girl," he says to his regular.

She stands up on her hind flippers and leans her sleek body and side flippers up against the table leg alongside Jose Luis. Balancing her head on the tabletop, whiskers quivering, Pancha's pool eyes are absorbing the fish being sliced into bits in front of her. I'm wondering if she'll grab for the fish less than a foot from her mouth and escape into the sea.

"Pancha never steals fish off our tables. She's been coming to me for almost a year now and is quite polite," Jose Luis tells me. Although wild pets are illegal in Galapagos, bonding when fishermen share their scraps is a silvery area of the law. What fisherman doesn't have a hungry crew flying or swimming around the back of his boat or at dockside as he cleans his catch? Visitors' cameras are snapping and locals buying dinner seem to enjoy watching their fish being ogled by a lovely sea lion.

One more fisherman offloads his boat, delivering a mess of the elongated metallic wahoo and bullet-shaped tuna onto another cleaning table. Pancha is alert and she knows Jose Luis is gone. In a flash, she lunges, and plunges off the seawall with her catch. Pancha is clearly not domesticated. Swearing in her wake, the fishermen watch her disappear. One predicts she'll be back tomorrow. Jose Luis will be happy, and like all the visitors, I will be, too.

Here's a book for your bedtime reading: *Lonesome George: The Life and Loves of a Conservation Icon* (Nichols 2006). Makes me remember stories around town of a tortoise expert, a sexy European woman who arrived in Puerto Ayora with the mission to extract Lonesome George's semen. But like the female tortoises put in the compound with George, she failed and came in for ribald joking in the scientific community in Puerto Ayora based at the Darwin Station.

Psyched for the upcoming Turtle Workshop in Cuba together, your video camera primed for our fishing village interviews.

Big pats to all the cats, and hugs to Fernando, too,

Sarah

ISABELEÑOS ANGLE TOWARD TOURISM

Coastal families have become more desperate than ever to sustain the consumer lifestyle they developed during boom times. With the pepino decline, fishermen strike in the hopes of keeping that lifestyle. Conservationists have been musing on how to cut Galapagos's human population, and especially how to get rid of fishermen. Trials of deep-water fisheries fail from the biological perspective as a possible alternative to pepinos; no fisheries substitutes have been identified by biologists.

Park officials feel they are reduced to seriously lowering the total number of licensed fishermen. Fishermen must maintain licenses, so by timing Isabela's fishermen's relicensing exams to overlap with the limited days of the 2007 pepino season, the Park successfully limits fishing effort. Because the exams are held locally to avoid travel costs, and at a greatly reduced fee, fishermen take the bait. Many decide to spend less time diving the scarcer pepino to reap immediate savings around the exam.

Brainstorming about how to entice the human inhabitants to leave, or the fishermen to quit, are common topics of conversation among the international scientists and conservationists in Galapagos. Some think in terms of a direct buyout to return residents to the continent; others think in terms of strict enforcement of migration and management measures to block (re)entry. One particularly inventive buyout fantasy, which I'm privy to over lunch, proposes offering each family, besides cash, an education package complete with U.S. green cards. Meanwhile, with births and newcomers escalating the population, management and mitigation costs could easily end up far dearer in the long run than even the more whimsical approaches being considered today.

Just as the exchange of fishing licenses for tourism licenses remains a tease, the fisheries co-ops ask the government to kick out the long-standing sport diving enterprises based in Puerto Ayora who are operating in good faith while the licensing procedure is up in the air. On tenterhooks, the sport diving operators counter that unlike themselves, the fishermen's culture of low safety standards would be dangerous for assuming tour operations that demand strict recreational dive safety, clear air filters, and other gear maintenance. Government finally sides with the dive operators and awards them long-awaited tourist licenses.

Fishermen, however, remain without tourist licenses, and would need much more than a paper permit to switch into tourism. The culture of safety is one reason given for their slow transition from fisheries into owning tour and sport diving businesses. It would be a cultural shift for fishermen to embrace regulations calling for advanced planning and backup systems. With subsidies to purchase backup and safety gear, fishermen would adapt instead of subjecting tourists to the high level of risks that they find acceptable for themselves.

Government and the Park are loath to award fishermen tourist licenses, although for several years, the government had been hinting at fishermen swapping for tourist licenses. In 2008, they finally present an elaborate point system to orchestrate the reduction of the number of fishermen. Points are earned by completing everything from environmental to community impact studies, which require hiring costly continental experts. Points must also come from retiring four boats that have fishing permits—*permiso de pesca*—or one of the few big boats that are permitted. Garnering all the points for a tourism license costs an exorbitant $50,000, out of reach for a single fisherman: four boat registration permits at $10,000 each plus $10,000 for the studies.

The co-op has roughly one hundred boats, each registered with a permiso de pesca. However, about sixty of these boats actually fish, and the rest are broken or unused, solely a fleet of ink floating on paper that can be bartered away to satisfy government requirements to retire boats to shift into tourism.

The co-op's professional director offers me numbers. There are 216 members (Gravez et al. 2008); each member is granted a fishing license—*parma de pesca*—a right that requires co-op membership. But nonfishermen have joined the co-op in recent years in the hope of availing themselves of the pending swap for tourist licenses. It's not a mystery in the community who is really a regular fisherman, who goes to sea seasonally or sporadically, and who is purely a paper waterman. Isabela's co-op leaders have been toying with only renewing fishermen licenses for those dedicated to fishing. Some think that by lowering the official number of fishermen, the Park would be pleased. Fishing competition would be reduced, as would competition down the road for tourism license swaps from their own faux fishermen members who have more economic facility to buy into tourism.

The co-op has held meetings to discuss the choices and consequences of this trade. Fishermen understand how the point system encourages them to band together to retire the four boats and to cover the cost of the studies. It's still acutely difficult to apply without an outside backer, someone with start-up capital and acumen for tourism. So far, five or six wealthy businessmen from Ecuador have invested in this process for local fishermen to gain the precious tourist license. An air of secrecy and shame prevails over becoming such a "face."

The point system for fishermen to shift into tourism rings hollow. Those fishermen who shift, as well as their progeny, can never return to fisheries, yet lack the training programs and expertise, subsidies, and funding to establish themselves. But, as a "face," the former fisherman probably is promised a monthly stipend for being the front for the one with these assets. The businessman, in turn, can earn $60,000 per month with a Galapagos tourist license, happily paying the "face" $1,000 per month. This practice would shift ownership to outsiders, and slowly undermine local tourism as birth-rights and hope for wealth are sold off.

Tourist license requests on Isabela are pending. If awarded, Villamil's number of boats with fishing permits would shrink by one-third. Roughly sixty boats would retire, leaving an active fleet of about forty boats. Because each remaining boat is crewed by three fishermen—a dive boat usually carrying two divers and one man running the compressor at the surface—120 fishermen could continue working at sea. This would accommodate the regular fishermen, and some seasonal fishermen. But it's all academic; there are no strong fisheries during the pepino ban.

Flights of Planning for Isabela's Tourism Boom

Ecuador's president actually visited Villamil in 2004 and promised to widen the tiny airstrip for jets, like those landing at the other two town's airports. Transportation infrastructure is a globally critical component for development, and Isabela's small airstrip only accommodates island-hopper planes within the Galapagos and military flights. Real estate prices start rising as Villamil poises for commercial flights from the continent.

After a financial debacle when the airstrip and communications tower were revamped inadequately and failed inspection, local rumors fly as to who profited. Reconstruction resumed on the airport building itself, which was completed by the mayor's office. Vendors buy up stall space at the new airport. Yet the mandatory components, such as the tower and runway, remain stalled. In January 2009, the mayor remarks that developing more slowly is better for the community, that it would be premature for large-scale planes to be landing tourists here. The architectural rendering with the jet on the runway, part of the still-stalled plan, hangs behind his desk. Island hoppers, however, increase their flights from several per week to two or three per day.

The alternative means of transport for most remains water taxis from Puerto Ayora that cost less than a third of the island hoppers, although the ride is for seaworthy souls. The run across open water takes almost two hours, contingent upon twin-engine power and sea and weather conditions. Water taxis busily ferry back and forth to Isabela in 2007–2008, as tour agencies in Puerto Ayora start sporting sidewalk billboards, touting Villamil as a quaint village destination, the perfect base for horseback-riding adventures up to the volcano and snorkeling with penguins. Locals and those on tours of duty no longer make up the majority of passengers. A few Isabela fishermen and a local bar owner originally converted their boats into a taxi service back in 2004. Competition mounts to carry interested tourists to Isabela. Villamil's dream tourism has begun.

The history of Galapagos tourism has been dominated by yacht tours—yachts include cruise ships in Galapagos tourism parlance. Since the late 1960s, the large-scale company Metropolitan Tours grabbed the concept and control of Galapagos tourism from Puerto Ayora's small-scale boat operators with land-based tourism from pioneers such as the Nelsons with Hotel Galapagos (Jack Nelson, personal communication). Until the millennium, over 90 percent of tourists in Galapagos slept aboard yachts, touring uninhabited sites. Public relations for yacht-based tourism still pitches the natural capital of Galapagos—the archipelago being listed as a UNESCO Natural World Heritage Site. Yacht tour guides rarely move beyond the topic of nature, nor do visitors query them. Tours have been structured as though the archipelago is void of human settlements. Among myriad tour companies, only the upscale ship of Lindblad had been stopping at Villamil, landing tourists for a quick visit twice a month. Isabela's uninhabited western shore, however, is a prime cruise destination. Extremely popular, the west coast provides glimpses of unique flightless cormorants, the penguins, and perhaps a night fire volcanic eruption on Fernandina Island—Galapagos's newest, most western hot spot as the archipelago actively shifts westward. It is doubtful whether tourists have noticed and discussed the pepino divers harvesting in Canal de Bolívar with their campsites along the shore.

Rhetoric continues to be based on the notion of Darwin's mythologized laboratory in an isolated archipelago. Conservationist fundraisers and tour operators alike, market the natural heritage as a Rousseau idyll, of an animal kingdom living without fear of humans. The giant tortoise is their signature creature, slow, innocent, and munching on Galapagos *Opuntia* cactus-trees' padlike "pancakes." The tortoise image epitomizes this exclusive peaceful place where a lucky few make the pilgrimage of a lifetime. Galapagos's nature is pitched as *Jaws*'s opposite, a safe haven to dive among sharks that normally scare people. Moreover, visitors expect not to encounter potentially dangerous "Others"—different people or cultures.

Reflecting low demand and low expectations prior to 2006, Villamil built few facilities at which to sleep and eat, except for two beachfront hotels owned and operated by educated foreigners who became residents. A Swiss woman, Dora, created Ballena Azul, a simple set of cabins in contrast to nearby Casa Marita. Casa Marita, to me, remains the loveliest hotel, despite a few other high-end additions. Its multistoried organic shapes in white stucco, trimmed in local cactus wood and lava is a good example of the hallmark architectural style of the European Galapageños prevalent in Puerto Ayora.

Villamil became the target for tourist development, being the only one of the three towns where there was still open land to build new hotels and private mansions. As a portent of the tourist boom in Villamil, Dora sells her hotel with its prime beachfront location to the owners of Puerto Ayora's posh Red Mangrove Inn, who promptly resurrect the place as upscale lodging.

Local women start by opening up a half dozen or so simple places. Many more women confide that they wish to add guest facilities onto their modest homes, but feel stymied by lack of capital. Over the next five years—up through 2011—the number of beds in Villamil multiplies five-fold. The mayor, with his vested belief in tourism, completes his own multistory stylish Hotel Volcan across from the beach. Technically, only Isabela residents are allowed to establish these hotels. Approximately 30 percent of the hotel owners are born here, while the rest have become residents through living here more than a decade or marrying locally, mirroring Villamil's population overall. Although rules are stricter for hotels than for houses, anything is possible. National elite families are putting up grand houses looking seaward: three of six mansions are owned by families who swoop into Villamil on their private jets. Though legally no hotels can have outsider owners, the face system prevails in land-based tourism as in cruise tourism.

Despite wishing for a tourist boom to provide income, many Villamil families resent the idea of losing their cherished isolation from outsiders. Their renegade status represents the font of their pride and social well-being. The prospect of growing into another Puerto Ayora remains odious to them. They know Puerto Ayora lost that village feeling of knowing everyone on the street decades ago.

Old-timers from Puerto Ayora such as Chris, a longtime resident and artist originally from California, comes over from Villamil occasionally for a quiet retreat at Casa Marita. She can feel the impending change and hopes Villamil can retain a bit of local ambience. Chris reminisces that Villamil looks and feels how she remembers Puerto Ayora did in the early 1970s, before tourist investors with deep pockets and experience developed Puerto Ayora into Galapagos's tourist capital. The tourist scene has generated the high influx of migrants and the apocryphal plague of 2,000 taxis parading around this town.

In sharp contrast to the large towns of Puerto Ayora and Puerto Baquerizo Moreno, Villamil's sand roads provide a beguiling detail. The sandy surface serves scant traffic, locals on bicycles and people walking everywhere. The municipality has preserved these roads in a conscious effort to nurture local tourism. They are perceived as an asset for natural beauty. Though some in Villamil would prefer the cleaner housekeeping option of pavement, the mayor's office has prevailed and carries out sand road renourishment throughout town. However, underneath the old-fashioned aesthetics lies an ecological subversion. The municipality has mined the sand from underneath their sand roads, including those seemingly endless few miles running along the beach out of town to the Wall of Tears. They have been replacing the underlying sand layer with rocks, topping it with a thin sandy coat. Where there's traffic in town, rocks peak through in places. But to date, the sand is reminiscent of the pioneer Galapagos days.

Rene Sangolquin, whom I call "Tarzan" for his consummate ability to interact with nature, has been showing me details of life underwater and in the highlands of Santa Cruz for a dozen years. Rene is perpetually playful, full of hyperenergy, a character always innovating, whether trying out his latest cross between a hookah rig and a scuba tank, pulling a three-wheeler buggy by motorcycle, or shooting video on novel treks with enthusiasm. He was the first to take visitors to places now part of the signature day-tour repertoire, whether by boat to the sea lion nursery and the shark Love Canal, or by land to the Santa Rosa agricultural area in the Santa Cruz highlands where the giant tortoises come to mate. As a horseman and tracker, he has innovated taking visitors camping on his family's land, even building a few cabins for those who prefer a roof after bonfires and tortoise watching.

Rene is a member of the small group of self-made small-scale Day Tour Boat Operators in Puerto Ayora who have banded together with Jack Nelson to work through the costly permitting processes. His life reflects the evolution of Santa Cruz tourism. Jumping onto the Villamil wharf after our ride over on his comfortable little snorkel/dive tour, we head to the main plaza for lunch. He sheds his pavement shoes and starts strolling barefoot, saying "I love how this feels. This is how I used to be able to walk around Puerto Ayora."

Villamil's Town Environment

Unlike Puerto Ayora, nobody in Villamil takes for granted an instantaneous ride. Most walk around town, or bike. A ride often requires calling and waiting 10 to 15 minutes. Yet it is a relief not to dodge between the taxis as in Puerto Ayora, where municipal licensing schemes have caused a snaking monster. Despite all rides in town only costing $1, newcomers eagerly sublet taxis from the owners, shouldering the $40-per-day lease on top of gas expenses. It seems a mystery how the driver can end up with a profit.

A glut of motor vehicles and noise pollution is a potential danger for Villamil. To their credit, Villamil's municipality hasn't granted themselves licenses to play this taxi endgame. Isabela's Transport Co-op already has about forty members, half of whom are also fishermen, and there are about ten "pirates" operating separately.

Since the millennium, Villamil has gained a steady supply of electricity and running sweet water. Satellite telephones are bridging Isabela's lack of communication. *Casa de llamadas*—shops offering phone booths—on Villamil's main plaza also provide clear continental and international calls. Internet finally reached Isabela in 2004, when the local chamber for tourism was more in the fantasy stage. Computer facilities for students and public Wi-Fi for aspiring small hoteliers and their addicted visitors are beginning to come online without having to purchase costly private systems.

Well-intentioned foreign projects wish to help Villamil be environmentally oriented. Galapagos, in general, is a regular recipient of environmental aid. The one for street lighting that avoids light pollution epitomizes all the good intentions as well as the flaws in many such projects. Designer street lamps matching the upscale, European Galapageño aesthetic—large glass globes with the top half silvered, set on naturally curvaceous tree trunks/limbs—block upward reflections. In some places, these lamps cast a lovely downward yellow glow, while in other locations darkness reigns. A number of the lamps have quickly blown out, with nobody able to repair them. Problematic in a classic aid-project way, the donation has come with dependence on outside technology and costly spare parts that locals cannot maintain without incurring debt.

Instead of donating practice infrastructure for the Villamil community, the United States Agency for International Development (USAID) has tended to advance Galapagos tourism—a national priority. They have funded the network of Park trails leading to lookout points, and historical and environmental sites, such as the infamous Wall of Tears and the iguana nesting site at lovely Playa de Amor. USAID has also invested in the extensive boardwalk trail leading through the wetlands behind town, connecting to the Darwin Station Tortoise Rearing facility. This route and the others are well used and enjoyed by locals for exercise walks and weekend outings. During the day, tour groups will continue to make the routes more congested.

The two main political groups in the municipality mistrust and malign each other over who has been best at retaining the ideal of local tourism development. Both sides strive to maintain the current sharp distinction between tranquil Villamil and bustling Puerto Ayora. They nurture an appealingly natural face to attract tourists. Both claim to want to contain outsider interests and foster local tourism, and both point fingers at the other for countering this end.

Municipal mayoral elections shift power between the two groups. The former mayor's group—anti-President Correa—accuses the current mayor's group, who backs Correa, of planning to open up tourism opportunities in Villamil to anyone. But the current mayor's group points to the past approval of Iguana Crossing Spa—the outsized hotel built into the wetlands.

Before the real estate boom began, a former Ecuadorean president's family purchased prime municipal land on the coast at the far end of town, alongside the wetlands boardwalk to the Tortoise Rearing Station. Supposedly, the ex-mayor sold this former president's close relative the necessary Galapagos residency for $100,000, allowing him to be the front man to develop this environmental horror, bypassing Isabela's building codes. The spa flies guests in on their private jet, functioning as a "destination resort" like the Galapagos yachts—providing all services from tours to food—so the guests do not use other tourism facilities.

In Villamil, the spa epitomizes national elite ownership of tourism. It adds the distinction of harming the fragile wetlands ecosystem, having contributed to filling in and drying out of flamingo habitat, driving the pink-feathered tourist attraction away. A second culprit has been the coastal road blocking the sea's extreme tidal flow into the wetlands. Thus, flamingoes that stay behind town in the salt flats are on the wane. A bitter town joke against the spa is that next the owner will market mud from the flamingo flats as an exclusive beauty treatment, symbolizing his mining of wealth at their expense. To redeem the loss of wetlands, the mayor tries placing a hose of running freshwater under the visitor flamingo lookout platform in town. It's a folk remedy for maintaining charismatic flamingoes until the rainy season replenishes the system.

His action is parallel to the independent conceit among major shrimp pond owners in the early days of Ecuador's shrimp aquaculture boom. They prided themselves, like hacienda owners, in creating personal systems instead of sharing science (Meltzoff and LiPuma 1986c). My favorite was the secret shrimp-feed formula that involved tossing a dead donkey into the shrimp pond.

After extensive Villamil wetland studies, scientists from the leading Ecuadorian scientific university, the Polytechnic University of Guayaquil—Escuela Superior Politécnica del Litoral (ESPOL)—offers free mitigation (Dr. Fernando Arcos, personal communication, 2008). Perhaps there is ego jousting, but the Park turns the university down, frustrating the scientists who are part of the same national network-cum-friendship circle of biologists as those serving the Park.

The extension of Villamil's wharf with national government funding goes forward full tilt as one more enhancement project by the mayor's office, until the Park freezes construction for a couple of months, suing over the legalities of mangrove cutting vs. simply retrimming. The mayor, in turn, sues the Park for mangrove cutting for another project. By 2008, an aesthetic and highly functional wharf complex is finally completed, just as pepino income ends and waves of tourists are picking up. The Park on Isabela is instructed to take a more direct role in identifying tourist sites and requiring Park-licensed guides, as on the two main tourist islands, Santa Cruz and San Cristobal.

The municipality's new wharf that greets visitors provides slips for local fishing boats also used for harbor and day tours along the coast, besides the multiplying numbers of water taxis and day-tour boats from Puerto Ayora. By 2012, boats bearing visitors cross from Puerto Ayora several times a day, sporting twin 175-horsepower engines, or faster, that reduce the trip to less than two hours. Travelers get cushioned seats protected from sun and spray for $40 each way, but dolphins coming in to ride the bow wake seem to have vanished from this route. The trip is in stark contrast to those from the turn of the millennium, when a normal crossing involved hitching a ride with a fisherman and bouncing in the bottom of his one-engine slow skiff, exposed to the elements for hours.

Tourism as Ecosystem Curse or Livelihood Lifesaver

To the Ecuadorean government's dismay, in June 2007 UNESCO placed Galapagos on the Endangered World Heritage Sites list because it is "threatened by invasive species, growing tourism and immigration. . . . The number of days spent by passengers of cruise ships has increased by 150 percent over the past 15 years" (UNESCO 2007).

President Correa has feared that loss of World Heritage status would rain awful publicity upon Galapagos, reducing their desirability as a lifetime dream destination. It becomes a national mission to get off that list, which happens in 2010 to critical world reports about UNESCO, as in the *Economist* (2010).

The archipelago's conservation record has been damaged by overfishing during the lobster and pepino booms when marine management could be curtailed by protests. Savor the paradox as international news reports show striking fishermen who succeed in pushing the national elite owners of the cruise industry into lobbying against marine management to keep the appealingly peaceful image of Galapagos. Fishermen's short-term antimanagement stance has led to the pepino ban, leaving Villamil's coastal community turning from fishing toward a local tourism boom as their next livelihood *salvavida*—lifesaver.

Reliance on tourism is not stable. Its precarious state comes from external factors beyond control, like how they view the sea itself. Nobody can control or predict disruptions caused by terrorism or global recessions, or sea-level rise and earthquakes. Inflating fuel costs deeply affect island life, from transportation to the cost of importing necessities. Fuel can drive up prices whether for fancy tours or low-end tourism such as the simple accommodations and services most Isabeleños could offer.

Nevertheless, Isabeleños see Galapagos nature's magnetism for attracting tourists and hope it will provide a livelihood option, one that might match lifestyle expectations acquired during boom-time fisheries. Locals have grown anxious over the bad omens of charismatic penguins disappearing for a year, and of far fewer flamingoes wading and feeding behind town, sweeping their arched bills on elongated pink necks back and forth in the mudflats. The loss of penguins and flamingoes can indicate harm to the ecosystem and symbolize natural charm evaporating.

If tourism development follows the pattern in Puerto Ayora, Isabeleños risk losing the chance of owning and running their own businesses. For example, this can happen if they allow themselves to be used as "faces" for outside interests with deep pockets and expertise who are eager to construct substantial facilities and reap the majority of tourism profits.

Isabeleños on the coast and up in the highlands pride themselves on leading an independent, relatively isolated existence, even as the gap is quickly narrowing through transportation and communication, and pending inundation by tourists. Locals still identify themselves as the Wild West renegades, outside many regulations, and they prefer their own municipal organizations that exist in parallel with provincial ones for health, education, and tourism.

CONCLUSIONS: TRANSFORMATION WITHIN A FRAGILE ECOSYSTEM

The small port of Villamil with its lava foundation of endless contours and the ubiquitous sea lions and marine iguanas, flamingoes, and penguins, still feels rural compared to the big sticks of Puerto Ayora, the bustling tourist capital on the island of Santa Cruz.

When I start coming to Galapagos around the turn of the millennium, Puerto Ayora has less than half its current population. There is only a fraction of the crowd of white pickup taxis cruising Puerto Ayora. I've watched both human and machine numbers surge into a scene where newcomers overwhelm the sense of being in an insular community, and I'm dashing out of the paths of taxis on the hunt for fares. On a weekend, families out for an ice cream and a stroll along the Puerto Ayora waterfront like to stop and chat, telling me how they love the pure air here, and how they experience a sense of peace. They feel it's a good safe place to raise children. Lacking memories of Puerto Ayora in its village days, they compare their positive sense of social well-being in Puerto Ayora to Guayaquil's stressfully dirty, jobless, crime-ridden streets.

Longtime Puerto Ayora residents, however, rue the loss of knowing practically everyone they pass on the streets. Most migrants landed in Puerto Ayora after the early 1970s inception of industrial tourism. This tourism development began drawing in thousands of migrants to work at wages often 75 percent higher than those on the continent. Among those drawn by the fisheries booms, a few migrated farther on to Villamil. Villamil still hovers at about 10 percent of Puerto Ayora's population.

Laws were designed in 1997 to sharply restrict new residents, but with scant political will for enforcement, the laws are paper tigers. Illegal immigration finally may be coming under some control by 2012. At this point, local common wisdom says that half the human population growth in Galapagos derives from births.

Puerto Ayora old-timers visit Villamil for a break. Walking barefoot along the sandy roads is something memorable and precious, a way of life that still exists in Villamil, a metaphor for their port forty years in the past. Puerto Ayora people relate to this place as caught in a time warp, reminding them of why they originally settled in Puerto Ayora. But they do not relate to Isabeleños and vice versa.

Isabeleños self-identify as being different from people living in Puerto Ayora and the other two populated islands of the archipelago. Residents feel a provincial pride in where they originated pre-Galapagos, and might still maintain family links on the continent. Yet they adopt a primary identity as Isabeleño, with the proud culture of being isolated independent renegades. The penal colony aspect is considered quaint history, and the finca foundation of local society has shifted down to the seaside with fisheries booms. But the pride in ruling their own lives is strong and similar to a fisherman's sense of freedom working at sea, self-employed, and braving nature. They support their municipality in organizing separate superior systems parallel to government services for health and education. Even for tourism the municipality created their own special Isabela chamber.

In the Villamil community where most Isabeleños have focused on fisheries booms—first in lobster and then in pepino—men united into their own Isabela fisheries cooperative. Leaders of the fisheries cooperative have held tight rein and sway, defending their community's extractive heritage against management limits. Legal and illegal distinctions have blurred the powerful sense of entitlement to resource use, their piece of the Galapagos economic pie.

Giant tortoises have served as the taboo provision in fishing camps out of sight from Park authorities on the far western shores of Isabela. Here divers kept hunting inside pepino's main nursery area of Canal de Bolívar. Galapagos fisheries leaders refused to give up this contested place to managers, and only the economic crash got fishermen to abandon their campsites and desert the canal so that it could recuperate.

Smuggling has become more difficult with the decreased isolation and increased vigilance caused by increasing tourism. This is a common pattern, one I also find along the coast of Chile, especially in Pisagua (Meltzoff 2001).

Villamil men who had been tempted by the Asian market for dried shark fins—especially before the pepino boom and after the money dried up—find finning has been sidelined to a few men, usually those in desperate financial situations or those who are the last renegades.

Fisheries co-op leaders and their striking followers did manage to curtail management through an essential alliance with the owners of yacht tourism. Successful in the short run, the Isabela coastal community faced pepino harvests tapering off before the economic crash. The women feared constant dive deaths as men dove deeper to find scarcer pepino. During the last couple of years of decline, Isabeleños and their municipality turned hopes and efforts toward grassroots tourism.

Eyeing the rest of Galapagos reaping wealth from visitors, Isabeleños are absolutely ready for a share of the tourism pie that had never before reached the periphery. But they want a tourism that belongs to them rather than to outsiders who buy up the beachfront and hire locals as workers, a common practice in so many coastal spots around the world.

Villamil's municipality has been working hard to welcome tourists, creating their own charming destination for land-based tourists interested in exploring Isabela for a few days. The now-famous sand roads circle the central plaza's artistically landscaped social space of lava rock and opuntia cactus trees. Aided by international agencies such as USAID and the World Wildlife Fund for Nature (WWF), the municipality now has scenic boardwalks and park-trail destinations strung outside of town along the coast to the Wall of Tears; up through the salt marsh to the Tortoise Rearing Station; and bordering the giant caldera to cross lava fields to Volcan Chico's lookout.

Real estate started to boom in the new millennium when the piper-sized airport was first being renovated to accommodate continental planes. The runway and control tower failed inspection and still don't qualify for flights that would transform tourist density overnight. That day will come, and land prices continue to rise in expectation. Meanwhile, boat taxis from Puerto Ayora are ferrying over unprecedented numbers for day tours or multiday packages.

Merry-Go-Round of Interests over the Galapagos Environmental Pie

Environmental interests have flipped. Now the Park is working in tandem with the owners of yacht tourism and their plethora of vessels against the interests of Isabeleño tourism. The municipality writes a letter to President Correa stating that the Park is contradicting conservation of the fragile ecosystem around Villamil. Even though the Park maintains the opposite, the inundation by yachts is going against the desires of the community and is not in the municipality's best interests.

EPILOGUE: DANGERS OF A CRUISE DELUGE

Villamil is in uproar in January 2011. Alliances are reshuffling. In an unconstitutionally secret move, the Park has signed an accord with the national elite's yacht tourism, allowing numerous vessels to start using Puerto Villamil as a stop on their Isabela itinerary. With Park consent that never went public, locals learn that the sixteen vessels approved for 2011 are slated to surge to sixty in one year. Pandemonium breaks loose.

In protest, locals forge alliances across their related tourism groups, from horse-tour operators, guides, and drivers to restaurant and hotel owners. All the local tourist interests converge in the central plaza in front of the municipality in loud public meetings to protest what they see as a serious invasion. They are upset at having Villamil's tranquility broken as their surrounding natural jewel is to be overrun by tourists brought in by outsiders.

The cruise companies have been eyeing Villamil's progress. For the first time, yachts in numbers will start anchoring in Puerto Villamil harbor, disgorging day visitors to roam the beaches and perhaps ride up around the volcano caldera. Locals argue that the yachts' barrage of tourists around town and nearby sites will cost the community its ambience without profiting their businesses. Yachts operate like destination resorts, providing all the services themselves. This confirms a local nightmare that Isabeleños would never be left alone to enjoy the biggest piece of their own tourism pie.

During the town uproar, I drop by to talk to Byron Gomez in his role as astute young vice mayor. He is definitely a community motor and understands various angles on issues. In the corner of the yard, his father, Jose Gomez, is stirring a caldron full of his handpicked prize green coffee beans over hot flames with half a weather-beaten oar as we sip his thick organic brew.

"If these yachts go unchecked, they'll harm Villamil's chances for tourism, and right when we're finally getting attention from the more adventurous visitors in Santa Cruz who are coming over for a few days. Public costs will definitely be higher than the pittance being offered the municipality. The Park granted them landing privileges without following the national constitution by consulting us. They're required to ask local permission, so we've gotten a tough lawyer and will go all the way to the president to stop them if necessary. Correa would help us!"

Back at the plaza, men with sly grins mention resurrecting strikes, their method of getting action. Alliances have flipped: fishermen no longer need elites in tourism to lobby to block fisheries management.

Park leadership heads over to these roiled waters and confronts Isabeleños, who have never been considered a gentle bunch of pushovers. In front of an emergency public meeting, the Park leader placates the crowd, demurring that perhaps Villamil isn't ready yet for this scale of tourism. He promises to postpone the pending sixty yachts and ban those with one hundred or more passengers.

The Park leader overlooks the point that Isabeleños want their place for their own tourism businesses, instead of being cluttered up by outsider interests. They do not want to be used as one more tour site for the yachts. Tours already have over fifty Park-approved nature destinations in the archipelago, based on protecting the animals and the safety of the visitors. Villamil is not necessary to cruise itineraries and is just an additional day tour, and perhaps a calm anchorage for the smaller vessels that might choose not to travel at night. But it is not a matter of needed comfort or safety; these smaller vessels have been cruising overnight for decades directly between the west coast of Isabela and a harbor such as Puerto Ayora.

The concomitant rise in trash, sewage, noise, party crowds, and security will have financial as well as social costs. Social well-being is at stake as much as control and ownership of tourism. Fieldwork in Key West (Meltzoff 1997) showed me the steep trade-offs from day-tripping cruise-ship tourism where promised community income never covers the financial costs, let alone the social ones. The atmospheric community has been nurtured by their municipality for the benefit of local tourism development to fund their social well-being and lifestyle without illegal or dangerous fisheries.

At the start of 2012, eight vessels per day pull into Villamil harbor for day tours. It's the beginning of the sixty vessels per week. They bring ashore noisy crowds, leaving behind beach and street trash, and sewage, and generally undermining the atmosphere around town so carefully groomed by the municipality to nurture Villamil's own land-based tourism. Ironically, Isabela has inadvertently lured in yacht tourism when they only meant to create a charming ambience for themselves by maintaining sand roads and redesigning the town's aesthetics.

LETTER: ECUADOREAN AMAZONIA WHERE OIL TRUMPS TOURISM

Hey Lila,

Hoping all is happy and tranquil for the holidays. I'll be missing you as my Galapagos TA when I head down to Patagonia for a drastically different climate of fieldwork. But first, I'm exploring Ecuadorean Amazonia from freshwater fishes to the indigenous hunters. I wanted to experience the Ecuadorean font of oil, the national wealth that trumps Galapagos tourism.

A week at a Rio Napo conservation field station flies through the shamanic full moon ritual with the Kichua's anaconda spirit on the silver river; night forest walk of gargantuan spiders and insects; and a day trip to a rescue foundation for confiscated wildlife pets from toucans, ocelots, and spider monkeys to a feral weasel cat and a caiman.

Friday is the weekly wood market on the Rio Napo where men balancing seven-foot timbers scale the tall, mud-slick riverbank. In early light, I slide down to the edge to talk to buyers and watch as more and more indigenous people, mainly Kichua, take turns pulling into the bank. They've floated the timber down tributaries bound into rafts, buoyed by balsa and bamboo, cutting the primary forest without replanting. The buyers, including three women from the cities, explain to me how they pay only a few dollars at this point of *primera venta*—first sale. It's hard to stop cutting and earning, but instead think ahead to when these trees could be gold in the forest bank.

Hiring a long canoe for the ride down Rio Napo, as much pulled along by the current as puttering, I spot oil spilling raw, spanning black out into the river from a broken tube. My boatman swings in closer so I can take pictures, until a manager runs over to his futile booms, shrugging to say "Give me a break!" then waving us away.

By the heat of the day, we land at Coca, the oilmen's eastern frontier. Their town reeks rough and tough, accommodating thirsty men. A pet toucan hops around my daypack as I eat rice and beans on the riverside veranda. The guy at the desk advises me to lie low or stick to very public places because of Colombian kidnappers who grudge the U.S. war on their drug traffic. We're in the frontier. Walking around the central market, the first stall has a headless armadillo rolled on its back, a pan of squirmy fat white grubs to fry up, and a large rodent leg. I begged off buying . . . no kitchen. But, I did negotiate a pair of rubber mud boots, a hot item all around town.

The Tiputini Biological Station guy fetches me at daybreak to travel further east where the forest still stands. We canoe a bit back up the Rio Napo to a head road, where a jeep bounces us over endless dirt tracks to another boat, to navigate the last miles along a narrow tributary inside the Yasuni Reserve. Here, Tiputini is set in 650 hectares of primary forest, including the flood forest that can be underwater from April through July. Universidad San Francisco de Quito, with Boston University, has a twenty-five-year government lease on Tiputini, and is petitioning the government for another twenty-five years. But, the current government is unwilling to step on the toes of the oil companies in the region who wish to end Tiputini's local mission of education and conservation research.

Late yesterday afternoon, the guide takes me piranha fishing for the first time. River dolphins pass us; then we enter into the mouth of a tinier blackwater river where the tannins meet the muddy milk waters of the main flow. Rapidly, piranhas ate up the red meat bait. Their wild red eyes shifted and they made urgent grunts as they posed, then the juveniles, or perhaps a smaller species, got tossed back. I'm glad to say they were my three. I never want to be the one that brings fish to their death, even though I love to eat them. City hypocrite.

On the way back floating downstream, river dolphins are coming up around us, sometimes two and three at once. It is nearing dusk and to top it off, a tapir starts swimming across the river in front of us. Called stinky cow or forest cow, or *danta*, she's large like a bovine, and what a snout and shiny dark brown wet fur. She makes a beeline for the opposite shore, plumb hips nimbly climbing the slope, and disappears into the jungle. . . . I don't take my eyes off her to photograph. I just stare.

One piranha makes it to our dinner table, a tough yet tasty fish. After piranha and chicken, I go pick up one of the teacups always waiting upside down on the drinks table in the native wood structure of the dining room. There is a strange mottled gray and brown leaf wedging in the one I pick up. I bring the cup up close to my face for a better view and almost reach inside to pluck it out. Somehow, instead I walk over to show the cup's content to Jaime, Tiputini's administrator and Ph.D. in bat zoology. He slowly holds out his hand for me to pass him the cup, walks a few steps to the waist-high leaf wall and shakes it out over the bushes below. OK, he´s a funny guy, so when he says I'm almost dead, I laugh, "Enserio?"

"Si, a young banana spider, a leaping type who bites with a lethal neurotoxin." My heart starts to pound as we sit back down around the dinner table. Jaime explains that if I had indeed stuck in my finger, or moved my nose a millimeter closer, we now would all be rushing by canoe in the dark, me in death throes. I feel extremely grateful for having shared the odd sighting. Isn't life like that hair's breadth away from death, with the illusion of solid permanence?

That night my mind focuses on how the spider could have gotten underneath the cup in the first place. I wonder if she was tired, having been locked up inside the cup all day. Days later I realize the banana spider could have easily lifted the cup up to crawl underneath, creating a shelter like with any small piece of wood littering the rain forest floor. The spider was incorporating cups into her environment.

Predawn today, I start back to Quito. The first leg stops at the Waorani indigenous people's community bridge. Nado, a Kichua man married to a Waorani, casts his net, pulling in all sorts of strange fish that he lets me examine . . . one with only one long ventral fin, another with an elongated yet blunt nose. Nado says he has joined a number of the four thousand or so Waorani left in existence who are settling around the oil company bases.

He tells me how oil executives in the thirty-five or so companies now helicopter directly into their isolated fortress bases deep in the forest, dropping down into the heart of their road webs dissecting the forest because of Colombians targeting these "gringos."

Some *petroleros*—oilmen—are buying Waorani passivity toward them in ways that drastically disrupt wildlife and cultural conservation. Here, for example, the petroleros arm Waorani with ammunition and rifles, then drive them on hunting parties. They kill enough howler monkeys in one day to fill the back of a truck. Gone is the time when they spent several days laboriously tracking, and shooting multiple times with the eight-foot blowguns I see around the community, to carry home what they can consume.

Waorani traditionally are headhunters, but the petroleros give them guns almost to encourage killing enemies and taking heads. Raids unwittingly defeat public sympathy for their land rights versus the oil companies. The press jumps on headhunting to justify even the wildest proposals to bomb the indigenous people out of the Amazon.

On to another kind of isolated frontier environment from Isabela, the inland seas of Patagonia where people have colonized islands in the cold rain in another kind of boom-time gold-rush fishery.

Soon from the land of volcanoes and glaciers,

Besitos mi hija,

Sarah

Chapter Three

Rain and Tears in the Fishing Camps

Merluza Boom to Bust in Patagonia

There is no way to happiness. Happiness is the way. —The Buddha

A NEWLYWED'S DISCOVERY OF THE MERLUZA MARKET

As a Chilean newlywed visiting the major Spanish fishing port of Las Palmas in the Canary Islands, Carlos Castaign dined on *merluza* (hake) steak and was stunned by the great prices this fish commanded for what was considered cheap fare back home. In southern Chile, merluza was caught on a small scale for subsistence, eaten fried, and the liver converted into a folk medicine substitute for cod liver oil. So Carlos Castaign and his family took advantage of the difference in market value and launched exporting fresh merluza to Spain in the early 1980s. They were the forerunners of what grew into a phenomenal boom that ultimately converted men eager for income into fishermen, and spurred their colonization of Patagonia's isolated islands in the inland sea.

It's the Patagonian summertime, wet and chilly, and I'm walking along the waterfront of Puerto Montt, by the major industrial city port and political heart of Region X (figure 3.1). Climbing up the narrow road behind Castaign's restaurant after a rich lunch, I reach the regional government headquarters. There's a commanding view of the bay. I'm interviewing officials and all levels of labor in the fisheries industry. Puerto Montt is the hometown of Castaign, who started the first merluza export company here. Down along the shoreline, a veritable hill of wood chips destined for Japan towers near the ferry offices selling tickets to journey southward into the fjords. At the far end of town, fishmongers shout, advertising their rare produce. Under their roofed seafood market, I wander among tables spilling over with seafood juices and a profusion of exotic creatures, some that I've never seen in all my days exploring fish markets from Japan and Oceania to northern Chile and Peru and to Spain and the Florida Keys (Meltzoff 1997). Next door in small food stalls by the water's edge, white-aproned hawkers compete at calling in customers for that day's offering. Choosing different spots each day, I sample fist-size picoroco barnacles fanning in and out of their jagged calciferous peaks, and try other local delicacies, such as the spongy, electric red piure with the aftertaste of iodine. Chilean-ranched salmon as well as merluza are also popular items on the menu.

THE BUREAUCRATIZATION OF CHILEAN FISHERIES

Perhaps 40 percent of the population in northern Chile lives off the ocean, between the artisanal and industrial sectors. Before the rise of industrial fishing, and the bureaucratization of Chilean fishing in general, the government cared little about delineating and defining artisanal fisheries. Artisanal fishermen were simply the ones who supplied food from the sea for themselves and locals. The government wasn't interested in regulating the artisanal sector, considering their resource extraction and economic contribution insignificant.

Ley Merino, established in the late 1980s, was the Pinochet era's forerunner to the 1991 Fisheries Law. A major difference between the two documents lay in the former's focus on managing the industrial fisheries, and the latter's shift to include the artisanal sector. The 1991 Law was born at a time when budding artisanal export markets were becoming economically significant. Remember the adage that fisheries exist because of markets, not the inverse. With the rise of industrial fisheries and artisanal exports, Chile has duly created managers who've spawned mounds of management red tape, though it is always slightly behind on-the-ground developments and out of synch with market booms, as we see in every case.

Fishermen must be registered with their corresponding caletas, and the National Fisheries Service keeps up-to-date detailed books on each caleta's operations. To qualify as artisanal, a fisherman's craft must be under fifty tons and less than eighteen meters long. At the National Fisheries Service headquarters in the capital, there are four main fishing techniques listed as artisanal, called *artes*—arts: (1) handlines; (2) gillnets; (3) traps; and (4) longlines. Caletas are listed and categorized by region, membership size, location of fishing grounds, and numbers and types of fish landed.

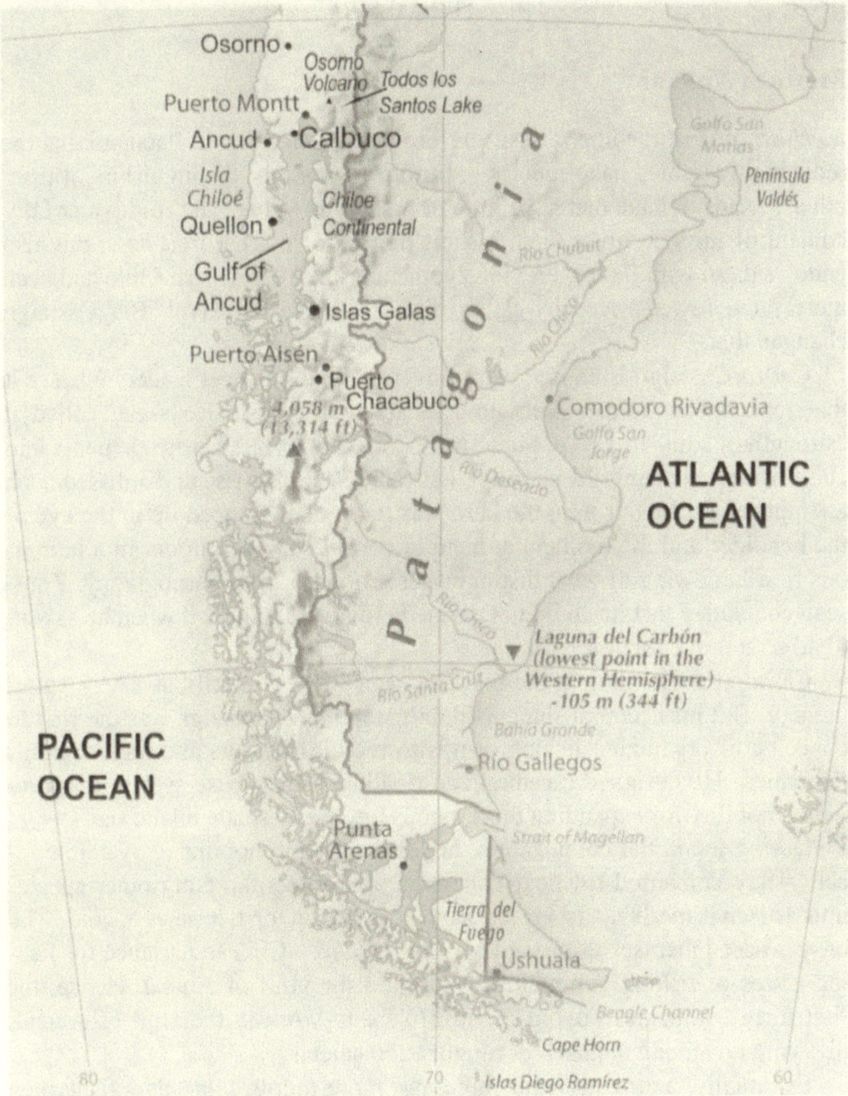

Figure 3.1. Patagonia map. Adapted from base map from https://www.cia.gov/library/publications/the-world-factbook/.

ETHNOGRAPHIC STORY AND EMERGENCE OF GLOBAL THEMES

Merluza Austral

Merluza austral (southern hake), referred to as *merluza* in Patagonia, is the most highly valued hake species in European markets, distinguishing it from other species of hake that have little or no market value. The cold-water fatty content of the fish gives them a short margin of market freshness, but also endows them with flavor. Previously, the merluza of southern Chile had been merely the lower-priced by-catch of *jurel*—horse mackerel. But Castaign changed that.

Carlos Castaign is an innovator, inventor, and business leader. What is it that makes one human seem to rise above the rest? Rousseau called it "strength of soul" in *Emile*. Modern wisdom categorizes heroic elements into classificatory schemes, descriptive lists, and such. Rousseau dismissed such attempts as futile. For him, the hero was too much wrapped up in the eye of the beholder and the moment at hand to permit easy definition. In a humorous foreshadowing of what distinguishes erotic art from pornography, Rousseau concluded that he could not define heroism, but knew it when he saw it. I had seen it in Castaign.

Castaign's triumphs in the merluza saga were primarily as savvy entrepreneur. But his accomplishments didn't stop there. Castaign was the first to conceive of organizing fishing camps to recruit migrants and train them as fishermen. He designed a centralized floating fishing-base system that provided mobility for expanding these camps into the desolate inland sea.

During the initial colonization, buyers began by mobilizing available locals. They converted rowboat fishermen and divers targeting other species into artisanal merluza fishermen hand longlining for the newest gold. The buyers based themselves around existing *caletas*—Chilean parlance for fishing coves or fishing communities—around the Gulf of Ancud. Hence, the first forty companies buying merluza went to work in the Gulf of Ancud, divvying up among themselves roughly 200 caletas.

Eventually, as the need for more and more merluza and thus fishermen accelerated, many men new to the sea were drawn to the promise of lucrative work and moved into the area. Castaign's own company, EXPROMAR, developed seven new caletas using his pioneering model of fishing camps called *campamentos*. These camps enlisted migrants into the merluza fisheries and trained them in the necessary fishing skills. Don Carlos, however, only saw photographs of his camps, remaining in Puerto Montt to run EXPROMAR with the family.

Previous Livelihoods around the Gulf of Ancud

Until the 1970s, Chiloe Island as well as Continental Chiloe—the opposing mainland embracing the Gulf of Ancud—were the agricultural horn of plenty, producing potatoes, beets, apples, sheep, milk, and cheese for both subsistence and shipment to Puerto Montt and Santiago. However, as the price of potatoes and beetroot dropped, Chiloe became a hard place to eke out a living. A number left, but some Chiloe campesinos—farm workers—turned to fisheries. With the influx of cash income from fishing, local agriculture slid into collapse as people purchased milk, cheese, apples, and potatoes, instead of growing their own. Moreover, urban migrants into the area did not share the concept of honoring others' crops. Instead robbery rose and there were few police to halt the thefts. Chiloe farmers could easily sell two big cows to buy a boat, so in this way many migrated to fishing camps to earn a different livelihood.

David, one of the ingenious buyers, meets me for coffee, grilled cheese sandwiches, and German apple cakes at a popular Puerto Montt café. He says, "In the very beginning, local fishermen tempted by the elevated price, set out in the Gulf in their four-meter wooden rowboats to try their hand at the latest market wonder—merluza. About 3,000 men were part-time farming and fishing in the Gulf of Ancud area near shore as well as close to the big ports of Puerto Montt and Calbuco. The first few companies went to the existing caletas to buy merluza. I did this myself, in fact. I figured out it was best to come over in my large motorboat and troll those little rowboats far out around the gulf from dawn to dusk. Then I could get a decent catch to buy."

David continues, "Around Puerto Montt, some locals were using the fishing technology or art of *pincho*—making a jerking action on a line with up to five hooks. Others were hookah divers, descending with long hoses connected to a compressor, or they snorkeled and waded in the shallows collecting shellfish and algae in the tidal zone. We wanted to retread all these fishermen to go for merluza. Soon some company reps headed north to the capital's port of Valparaiso to recruit teachers from their two famous caletas."

I ask him what they were promised, how many came down, what they brought with them, and what they were taught. David replies, "They contracted 130 *porteños*, professional fishermen. These new 'professors' trailered their six-meter motorboats down on trucks, preparing mentally for our fiercer southern seas. Their families followed." The porteños set about teaching longlining, their specialty, their fishing art. For most gulf fishermen, this type of artisanal longlining was novel. It takes great strength to pull up the line by hand, heavy with the weight of 500 hooks, each hand-baited with a sardine that hopefully has been swallowed by a much grander merluza. Plus, preparing the baited hooks is a tedious job that's imperative. The

Porteño women frequently do it onshore at the caleta. This untangling and ordering of hooks also had to be learned and absorbed as part of the preparatory work (figure 3.2).

"What's life like there now?" I ask. "I'd love to go visit your fishermen friends."

David agrees, and we make plans for the following week. We board a bus and then ferry hop, heading along the Chiloe mainland to the *caleta* where David customarily docks, to meet his fishermen in their tiny boats and then to tow them around the gulf longlining merluza. We finally end up a short distance from a large wooden house—a construction in progress—owned by the *caleta* at the end of this elongated cove. The yard is bumpy and covered in dense grass growing in moist humps. It's a damp environment.

A young woman, Amanda, is tending her pot-belly stove to brew the *yerba* maté tea. Maté is the stimulating ritual social drink enjoyed throughout Patagonia, all of Argentina, and Uruguay. Amanda's thick blond hair is braided back and she's wearing one of this cold region's thick woolen sweaters that she knitted herself in rich sea blues. She leaves the kettle to boil and leads me to the back door where she points out the dirt road. "That carries on into the alerce rain forest." Alerce (*Fitzroya cupressoides)* grow for a thousand years and are the Patagonian equivalent of redwoods. Amanda senses how much I want to go see them and adds, "Roads are too muddy now and

Figure 3.2. Catching a juvenile merluza. Courtesy of the author.

even from the road the giants rest a long trek inland. Now is the season to travel by boat." Sigh.

A small group gathers to pass around the hand-sized gourd full of maté that is refilled with water and sugar after each person takes their turn downing the potent sweet tea through the silver-plated straw—*bombilla.* I love maté and its high energy, social and physical. It signals time for relaxing and sharing stories, as the meat grills on the fire outside the doorway. The party begins.

During the ferry ride along the green coast, David had described how the merluza fishery evolved. "Our new fishing companies never had to advertise. The unemployed flowed to us through word of mouth, desperate for any opportunity, knocking on office doors to sign up for the camps. Remember, the early 1980s economic crisis had people on the move all over Chile. They were clueless about the harsh realities they were signing on for. Urban youth who'd never even done manual labor arrived from as far away as central Chile and the capital, though most came from the big towns in between, such as Puerto Montt, Puerto Varas, Calbuco, and Osorno."

When I had asked David how the leap was made from caletas like the one we are going to visit to floating fishing bases, his eyes flashed. "Eventually, the need to expand and increase efficiency led Don Castaign to radically rethink who, how, and where. He invented the heart of the operation, a centralized floating fishing base—*la balsa.* Buyers would build one and then post a few youths as managers in the new place to set up a company store—*pulpería.* These boys were responsible for paper work, accounting and receiving fish, as well as reordering ice and supplies to keep life bearable."

This floating base system was able to provide the necessary mobility for expanding camps down into the distant reaches of the web of canals and rocky islets making up the inland sea. By making use of the floating base system, companies were able to increase fishing effort by moving deeper south into the nether lands of inhospitable Region XI, discovering new fishing grounds, and establishing camps along hitherto uninhabited fingers of water.

I asked when the women came, and David replied, "For nearly five years, women and families stayed away. Gradually, they began to arrive, integrating into the impromptu and incommodious camp life. They kept the men company and helped out, serving as crew when need be, and as baiters." David laughed, "Actually, fishermen hadn't left their women much choice, being unwilling to take time out of boom fishing for a visit home. They were making fortunes." Income was topping out at what was considered tremendous earnings. And so, home came to them, transforming camps into real communities. Castaign's system was responsible for organizing and galvanizing men into artisanal fishermen in this potent boom-time production mode.

The Makings of the Merluza Boom

In the beginning of the boom in 1981, the Spanish market absorbed all the Chilean merluza, and Castaign was having trouble finding enough fishermen to keep his Spanish buyer, Larrea, supplied. And Castaign's company, EXPROMAR, had to truck fish the long distance up to Santiago, as there was not enough initial volume to warrant air cargo flights from Puerto Montt. These fish sat, eyes glazing over, and had to have their heads cut off. They were not fresh enough—clouded eyes and dull red flesh—to pass basic muster by Spanish housewives. With years of Spanish fieldwork in the 1980s around the major Galician fishing port of Vigo (Meltzoff and LiPuma 1986a), I'd witnessed how the Spanish pride themselves on being seafood connoisseurs. Average consumers in restaurants and markets alike demanded full freshness, willing to pay top dollar for it, unlike American counterparts.

Larrea, who was from a Basque fishing family in northeast Spain and considered tough and clever by the Chileans, soon flew over to Chile himself, backed by capital from a Basque fishing company. He launched the second merluza export company. Soon, more and more small-scale buyers sought direct access to Chile's merluza and started their own companies in Puerto Montt. Now, to establish a better, bigger, and more stable export market, these buyers were in critical need of more merluza.

By 1983–1984, industrial fishing companies had begun to target and export the precious and abundant merluza, following closely on the heels of smaller buyers. Legally, the fleet of industrial trawlers (vessels twenty meters and greater) is limited to fishing in the ocean—referred to as the "exterior waters." In this way, the industry lacks direct access to the rich concentrations of juvenile and adult merluza inside the numerous fjords and finger canals. Because of these regulations, the Puerto Chacabuco plants turned to artisanal fishermen who were reaping these "interior waters," buying up their artisanal catch. The industry also was dictating price. During Chile's economic crisis of the early 1980s, Pesca Nova, the giant Galician fishing company of Vigo, Spain, was operating their fleet here as Pesca Chile, having successfully petitioned the sub-secretary of fisheries to buy industrial fishery licenses for merluza in Region X.

In a twist of fieldwork fate, during this time I was doing fisheries ethnography in Vigo from the Galician perspective, exploring Spain's entry into the European Union (EU) and the increased need for her fleets to acquire foreign fishing licenses to survive a climate of Exclusive Economic Zones (EEZ). Spain's fleets were now being feared as the Spanish Armada by EU fishing fleets. Chilean licenses were critical.

As the boom escalated on the artisanal fronts, Chile's government development agency, Corporación del Fomento de la Producción (CORFO), began its credit program for artisanal boats to alleviate the economic crisis and help encourage fishing efforts in Regions VII, X, and XI. Companies also lured potential fishermen by lending them money against the sale of future catches. Accordingly, orders shot up for the seven-meter boats with 13-horsepower Chinese inboards. Puerto Montt boat builders began briskly sawing away, building not only government-order artisanal boats but also the twelve- to fifteen-meter vessels used to transport the fresh merluza to Puerto Montt. At the height of the boom, an artisanal fisherman with one or two months' earnings could buy his own wooden boat with that inboard motor, including the necessary hooks and longline.

No one knows just how many people were actually fishing at the height of the boom, and it is just as unclear how many merluza were really caught. Governments globally recognize that census data for artisanal fishermen are frequently inaccurate. Here especially, the camps often formed spontaneously, outside of official channels. An artisanal fisherman is supposed to carry an identity card—*carnet*—that states "fisherman," and be listed in the Regional Fisheries registry. Officials guessed that perhaps 40 percent of all fishermen were registered. Yet, 60 percent were neither registered nor carried the carnet, preferring to avoid a trip to government offices and the cost of a photo. Instead, when asked for ID, these men show certificates from military service and schools.

Artisanal fishermen, in line with a set of worldwide patterns, often ignore or bend regulations. Responding in direct correlation to what buyers are willing to purchase and at what price, they shift the composition of landings to exclude or include juveniles. Meanwhile, fishery managers inadvertently, or sometimes even consciously, furnish catch landing statistics as accurate representations of in-the-water reality. Taken together with a poor fishermen census, managers work with data that are sketchy or questionable at best.

What is clear during the boom is that local and migrant artisanal fishermen were making fortunes in the merluza market, as were buyers. Peak catches coincided with peaks in the number of buyers and fishermen. By 1987, there were fifty to sixty companies in Region X. Each caleta might have two or three companies competing for its merluza, thus tripling 1984 prices. Intense demand was reflected by great prices—1,500 pesos per kilogram. Landings shot up at the fastest pace seen in any Chilean fishery at that time. Daily, 100 tons of high-quality fresh merluza could now be flown to Santiago bound for Madrid because the Puerto Montt exporters had gained a solid foothold in that major international market. At the boom's apex in 1988, two fishermen in a boat in the Gulf of Ancud could haul up 200 kilograms, while those in the inland seas to the south could fill their open boats with five times that amount. Fishing capacity increased as inland sea fishermen invested in bigger boats, some expanding up to the artisanal legal limit of eighteen meters. The motor alone for such boats costs $30,000; and their fishing technology includes a costly industrial hydraulic winch to haul up the longline. Pushing exploration further southward into open new grounds, buyers and their camp fishermen began migrating toward Aisen, the vital merluza spawning area.

Pinochet and Fisheries Policies

If ever a tale had an evil antagonist, the story of Chile has one in General Pinochet. His faction seized power via a military coup in 1973, overthrowing the elected President Allende. Years of violent and often murderous repression followed. As political shifts are inevitably reflected in changing fishery and other resource management policies, so is Pinochet unavoidably part of this story. A majority of industrial fishery owners supported Pinochet, while the fishermen themselves did not. The middle class in general backed the autocratic law and order, and the focus on export market growth promoted by the Chicago Boys running Chilean economic policy. Pinochet kept some of his promises, holding the 1989 Yes or No referendum where he was shocked to lose, stepping down from the presidency as a result. But the country remained divided because Pinochet retained power in the military sector. Owners and workers avoided political discussions for fear of retribution. In 1998, after a Spanish judge sought his extradition for crimes against Spanish citizens, Pinochet was arrested in England, and the strict limits on political dissent, finally relaxed. During these years of my fieldwork, I witnessed the spiraling process of workers, including fishermen, feeling free enough to speak out more publicly against the former dictator.

The incipient global era of EEZs coincided with the development of an industrial merluza fishery in Chile, bringing high-seas politics into play. With Chile's new control of the first 200 miles of their EEZ, foreign fishing concerns such as Pesca Nova of Vigo, Spain, sought and gained access to those waters from the government to fish merluza.

During the early 1980s, Chile's neoliberal economists were cultivating export markets while facilitating foreign investment. Castaign's efforts to develop merluza caught the eye of Pinochet's minister of economy, who personally jumped into the boom by starting his own private merluza export company. They were implementing policies to encourage export in a time of national economic crisis that had left millions jobless. The severe recession in the early 1980s created a sea of unemployed eager to fish under harsh conditions in the fjords of southern Chile. It was easy to trigger migration down into remote inland sea islands. At such a perfect time to man a boom, even those accustomed to urban amenities set out to pioneer cold, wet wilderness, persuaded by the promise of providential profits.

The government enhanced opportunities presented by the strong international merluza market through a national credit program for artisanal boat buying. Buyers were quick to take advantage, working with the existing fishing industry, with its fleet and shore plants, to help obtain artisanal boats via this government credit program.

"Most companies soon failed because they had no idea of fishermen," said Carlos Castaign, shaking his head at the lack of understanding. Companies right and left failed to foster a successful working relationship. Nonetheless, fishing effort soared. The camps grew, signing on and training migrant fishermen, campesinos, and jobless urban men. The first camps in Region X were largely male back in 1985. But by 1986 or 1987 whole families were arriving. They camped on myriad islands in *rucos*—crude dwellings, patching together scraps of canvas, plastic, wood, and tin. Many camps gradually evolved into fishing villages, building up churches, stores, clinics, and schools.

LETTER: JOURNEY TO ISLAS GALAS

Dear Lanny,

Soon I start the trip south through the inland sea of Patagonia to the Islas Galas, to an island named Isla Toto. I'm going without knowing anyone there but am sure I'll make friends—especially in a place named like Dorothy's dog! My goal is to go out fishing during the month's three legal days for catching merluza austral—giant tasty southern hake that the Spanish market craves. Locals ignore the weather, which can be fearsome, so I've brought a bright blue plastic slicker to wear over my techie, permeable-breathable Patagonia model.

From Galas, I'll need to hitch the next ride in a buyer's launch full of merluza bound for Puerto Chacabuco, the industrial port some sixteen hours south. Luis is an industrial hake manager, a mane of black hair and tall, whom I first met when he worked for the Puerto Montt regional government office, negotiating hake quotas. After yesterday's interview in Puerto Montt, Luis gave me his cell number so we can link up in Chacabuco, itself, over the merluza he must buy and fly to Madrid's fresh market. Luis is very open and knowledgeable, and luckily for me he's interested in thinking about and sharing different perspectives.

Sunday is a rare, clear, deep-blue day in this rainy climate, so I take the bus from Puerto Montt to visit my friends, the Family Kuschel at Vicente Perez Rosales National Park. They sit on Emerald Lake—Lago Todos los Santos—that glitters underneath Volcán Osorno's Fujiyama-perfect snowy peak. You would love taking your quirky marvelous photographs.

The Kuschels are one of nineteen families who work as fishing guides around this celebrated body of water bordering Argentina. The Kuschels are particularly fortunate because they live right on lakeside land in a wooden house where they've added on several rooms as a hostel. When their family first arrived in 1911, there was no US$200-per-night elaborate lodge opposite them, a few hundred feet across a fishing stream running down through steep falls cascading out of the lake.

Over the years, about three hundred wealthy families have surrounded the lake, building private faux castles miles apart, overlooking the spectacular volcanoes ringing the lake's emerald depths. These families own clear titles. However, the nineteen families of the fishing guides are fighting not to be evicted. Their leader is Kuschel's elder son, who travels into Puerto Montt periodically to see the regional *intendente* (manager). They must wrestle the national park authorities to be grandfathered into their long-standing settlement near the falls. The authorities drag on and on, hesitating because the owners of the lodge covet the Kuschel's site. It's a perfect case of tourism tensions highlighting class interactions between the large-scale owners and the workers.

The lodge wishes to expand to build a yacht club here, besides wanting to terminate this family-style accommodation complete with home cooking. The Kuschel mother welcomes me back with a warm hug saying, "Has it really been two years?" As she shows me to a little wooden room, she asks, "How about some of my smoked trout for lunch?"-

I'd already caught the aroma of baking bread. I'd watched her kids before, casting for trout right out in front. On my last visit, I'd gone to the village political meeting with her son, Roberto, where they'd strategized how to keep their homesteads. I'd given him an introduction to Luis, who was still in government, and back in Puerto Montt, gone and talked to others at the intendente's headquarters. I wonder if he followed through, but Roberto's off guiding now.

His parents catch me up over coffee around the pot-belly stove, the father saying, "Last year, we invited the intendente and other officials to a big trout bake here. They came, and Roberto gave the speech to everyone, pleading our case. But look, a year later, and there's still no action one way or the other!"

So the Kuschels and the others continue to live on the lake, and we can too as their guests. The next day, I share lunch with a fellow lodger, a Barcelona guy who's about to climb Volcán Osorno. He points up at the 2,630-meter peak saying, "Most climbers snowboard down. Me too. Look, you can see the ziggy marks running down." You can notice their trails, yet I'm thinking about these lake fishermen and their inland sea cousins, both hurtling along political slopes.

Happy to see you so soon in the holiday times and share photography.

Hugs to you and Susana,

XOXO Sarah

The Merluza Crash

In the wake of the merluza boom, communities in the caletas around Toto and elsewhere have struggled to redefine their niche, needing to expand into additional fisheries to survive. The merluza fishery peaked and ultimately crashed inside one decade, from 1982 to 1989.

In late 1988, all catches dropped precipitously to virtually nothing per boat, and the juveniles from the canals were pint sized. The bounty of merluza—a species that lives over thirty years, maturing slowly compared to other fishes—then appeared to disappear altogether in 1989. Don Carlos Castaign, like the minister of economy, sold his company EXPROMAR in 1989. A family of bankers from the capital of Santiago bought Castaign's "gold mine," but even with their deep pockets they could not hang on during the continued poor years of the crash. Both the artisanal and industrial sectors blamed each other for this startling stock failure. Each interest group believed the other caused the collapse by excessive fishing.

The industry accused the artisanal sector of injuring the merluza stock through excessive capture of juveniles within the inland sea, legally denoted as "interior waters," where only artisanal fisheries are allowed. Migrants had flooded the fishing camps that spread down along once-uninhabited canals, creating ever-new fishing grounds. Industrial as well as small-scale fresh fish buyers, however, profited from interior waters.

The industry, after encouraging the migration of fishermen into new areas and buying their catch, started to regard this push as detrimental to their own interests. They publicly rued having helped to establish bases for intensive artisanal fishing where their own large vessels were forbidden. Now industry was claiming that immature merluza were too few to biologically replenish the offshore stock. But a strategy to colonize wilderness waterways gave the industrial processing plants access to the rich stocks of merluza in the inland sea, an area otherwise off-limits to their sector. Industrial vessels only had fishing rights offshore, in what are legally termed "exterior waters."

Meanwhile, Galas fishermen point to the industry's high-tech solution to escalate landings. Their artisanal merluza association leader notes the disparity to me, "In one day, just two industrial trawlers catch the equivalent tonnage of 400 of our artisanal boats."

Stock assessors offer an original explanation: although biomass caught may not be greater, the mortality (number of individuals killed) in interior waters where artisanal fishermen longline is ten times higher than in the industrial fleet's exterior waters. Therefore, the industrial fleet could get the backing of these stock assessors to argue scientifically that artisanal fishermen neither recognize their own destructiveness nor understand the distinction between mortality and biomass, referring to numbers of fish as opposed to weight.

What nobody was considering, according to Dr. Nelson Ehrhardt—a Chilean fisheries biologist, stock assessor, and colleague at Rosenstiel School of Marine and Atmospheric Science, University of Miami, was that the collapse might also be a cyclical result of El Niño Southern Oscillation (personal communication, 2000). Massive increases in surface water temperatures for months to years at a time often result in marine life mortality and changes in migration pattern. Merluza can live over thirty years, slowly reaching reproductive age at nine years old. This is also the age at which merluza are "recruited" into the fishery—attaining legal catch size. In Ehrhardt's provocative "hole" scenario, during a major El Niño event, older merluza might change migration patterns and head for cooler waters by going deeper or heading to higher latitudes, but the young fry—one- to two-year-olds behaviorally bound to stay in shallower surface and coastal waters—would be killed by the effects of abnormally high water temperatures. Yet, because it takes so many years for merluza to mature and become legal catch size, the small fry would not immediately be missed by humans. Thus the "hole"—significant dearth of one- and two-year-old merluza—caused by the powerful 1982–1983 El Niño, would only be noticed after eight or nine years, exactly when their class should have been recruiting to the merluza fishery. The fishery actually collapsed on target according to Ehrhardt's "hole" theory, exactly nine years after this major event carved a "hole" in the resource. Timing of the collapse between 1989 and 1992 perfectly fits Ehrhardt's "hole" theory. It can also explain why *merluza* catches revived in 1993, once *merluza* born after the strong El Niño were coming of age.

Although overfishing is often the finger-pointing diagnosis for a fishery collapse, like in the case of the merluza, El Niño's cumulative ecological effects are easily root causes. Natural phenomena play powerful and still unknown roles in fish stocks. However, global management efforts tend to blame fishermen and then buyers as the most apparent causes of market plentitude and availability, or their opposites. Interactions are far more complex.

Isla Toto's Women

On my second day with Ladiseo and Gloria, I awake to pancakes and blue light vibrating between sky and sea. It's a rare sunny day, so I follow the wide wooden boardwalks interconnecting homes and bridging short spans between islets. Locals will be building more of these connecting structures with student summer volunteers from overseas churches. Strolling the boardwalks, admiring views of Galas's deep, ringed harbor, stopping to talk to people in tiny housefront shops, I end up where a mother and father are constructing something on the front of their boat as their kids splash in the transparent shallows nearby.

"We're building a bow water break so we can sleep onboard," the mother says, "Then we won't be forced to find a campsite for several days when we're at good fishing grounds."

"Boats keep fishing right up until the moment we're called back when the merluza quota is full. There's a communication delay in favor of a bit more fishing," says the father, smiling.

Recalling my fisheries fieldwork in other ports of the world—Chile, Peru, and Ecuador included—I know women are hardly welcome fishing at sea. To my excitement, I rapidly realize that in Galas, some couples fish together. In scant places, and under special circumstances where there's a pressing shortage of crew combined with a customary fishery unhinged from strict spiritual restrictions against women, I have seen women go to sea. In Galas, this might be partly attributed to pioneer partnership of sharing hardships within a tiny population base, filling in for a lack of crew, and needing to keep profits inside the family.

Among the saltwater Lau, men going turtle hunting far at sea off Fanalei, Small Malaita, I witnessed this once. In their own cultural hunts—such as for porpoise—the women taboo was strictly adhered to. But for spearing turtles, an adopted deep-sea hunt, they blurred the spiritual demands against women and occasionally took an unmarried daughter if they were short a male crew member. I couldn't get them to take me for anything. Women's menstrual blood in Malaita and in many cultures is seen as harmfully potent against men and their spirits. Thus, women on fishing boats antagonize helpful sea spirits. The fear of women's procreative forces, as I see it, is often flipped in language so that women are called "polluted" or "polluting," so male domains must be kept free and clear, and clean of their presence at sea. The taboo can even extend to where sacred fishing canoes and gear are stored, like sixteen kilometers down the coast from Fanalei, among the Sa'a tuna hunters (Meltzoff and LiPuma 1986b). In the Pacific, this fear of women extends to many other arenas for hunting, warfare, and health. Where I worked in Galicia, the Christian crews called women "bad luck" and refused to take any females on board—women were seen as a type of Jonah. An old Hells Angels initiation in New York City that was as risky and scary as they could imagine was oral sex during menstruation.

The highest likelihood of finding women fishing on boats is in enclosed waters, fishing in inland seas, lakes, and canals, however vast and wild protected water can become. When it's not the ocean, there is a chance that local concepts allow women to be viewed as capable. Two personal fieldwork examples jump to mind where women paddle out to fish without men. One is inside vast Pearl Lagoon in Nicaragua, where Garifuna women fish from canoes, and the other is inside the passages of Malaita, where Lau saltwater women paddle to trade fish and porpoise meat as well as to reef glean.

That women fish in the inland sea around Galas—albeit crewing for their men folk—is a real find. Sudden storms and tricky navigation in these frigid waters make it dangerously rigorous. There are some women throughout this book who harvest marine products: the sole young unmarried diver in Puerto Aldea, the women waders in Tongoy Bay and around Coquimbo, Region IV, and the seaweed harvesting women like Doña Lasteniain Caleta Caramucho south of Iquique, Region I. Nowhere in Chile or Peru have I encountered the phenomenon of women fishing from boats. In the seaward fishing communities, if women get involved in fisheries, they classically harvest on foot, wading near shore to glean tide pools, line fishing, or beach combing seaweed. Or, they help onshore by baiting hooks and preparing gear. Unlike their Galician counterparts, where women predominate selling their men folk's artisanal catch, women in fishing families here tend not to market the catch, although in Peru this is common.

On the Galas boardwalks it feels like summer vacation. Shouts from boys playing basketball echo from inside the gymnasium/pool recreation center. The center sits next to the boardwalk, also up on stilts wedged into the rocks on steep terrain. I wander over to another cove to join a group of little kids and a furry dog playing at fishing. Three of the kids tell me how they sometimes get to go out with their parents. If the weather seems clement—as it can be in summer—they fish merluza as a family. Mina and Mari, the two girls, are eight and nine, and Tino, the boy, is eleven.

He says, "It's really hard, baiting hooks on six longlines, tossing meters and meters of weighted line over board, then pulling them up arm over arm into the boat," demonstrating the arm action. "All kinds of fish besides merluza are coming up, like dog sharks and odd big-eyed fish that live way down." The family crew must work in a steady constant rhythm to stay competitive during the few days of the quota. I nod.

Mina adds, "When you pull fish into the boat, some of them have burst eyes even though some are wiggling. We have to unhook every one and throw back all but merluza."

Still a small boy, Tino, who is crouched over the tidal pool nearby, has been successful casting with a forged hook he'd baited with a worm. He swings in one of the fingerlings on his line, and another boy helps him add it to a plastic jar. Tallying, the total is ten so far. The dog in parallel play (I like this expression from my days studying child development and researching play) is chasing seabirds at the water's edge. She reminds me of Kim and my Mango dawg—an Aussie Shepherd whose photo I always carry with me—wheeling 360s, long legs and wolf tail spinning in circles, ecstatic concentration, sheering close over, over, and over in pure joy.

As the light lengthens and it's time to head home, Tino holds up an eye. Most of the jarred fish are already belly up from rough handling. I suggest letting them go swim back in the pool to grow up to be big enough to eat. Am

I silly or what, I realize, as the boy casts me a doubtful look. This is a game, forget it, they are just having fun playing go fish. We walk up toward the dog's house, and a mother opens the door to let in her hungry happy ones.

Neoliberal Fisheries Policy Development

Merluza is a great example of spontaneous, unregulated growth in the context of the rampant privatization and retooling for export that characterized the 1980s Chilean economy. At the onset of industrial fishing, besides the Spanish, the Japanese arrived with factory ships to target merluza. Chile exported much of the frozen industrial catch to Japan, while the fresh market was focused in Spain. By the late 1990s, the government agreed with the fishing sectors that overfishing had been the main cause of the merluza crash. The government fisheries biology investigation agency—IFOP (Instituto de Fomento Pesquero)—under marine biologist Patricio Bernal, had conducted a study showing that merluza were highly susceptible to overfishing in the inland sea. IFOP gradually accepted the idea that the merluza business had suffered badly from allowing fishing of interior water stocks that include many juveniles, who then could no longer replenish offshore stocks. When Bernal became the head of Chilean fisheries—sub-secretary of fisheries—he put his study into policy, setting up the quota system in 1994.

But as is typical with a spontaneous fisheries boom generated around a new market, there had been ample lag time before the government introduced any regulations. The National Fisheries Service had managed to put inspectors on the industrial boats at the rise of the frozen merluza industry, but they were inexperienced university youth who were told anything by the captains. The collapse began in 1989, but it took the new Fisheries Law from 1990 to 1993 to develop the legal mechanisms for creating a management system. Not until 1994 did the Chilean government through its National Marine Fisheries Service introduce the politics of quotas (*política de quotas*), an official quota system for merluza. Ironically, by that time the merluza fisheries had boomed, crashed, and were already rebounding. Quotas were based on the crash, but merluza were recuperating on their own, probably for the reasons put forth in Ehrhardt's hole theory.

As the new Fisheries Law's teeth were being honed, the industrial sector moved as one, lifting their entire operation further south into Puerto Chacabuco, Aisen. Spain's pair of industrial fishing giants, operating in Chile as Pesca Chile and Frío Sur, led the way, abandoning Puerto Montt and moving all forty to fifty of their twenty-meter trawlers. Despite the far colder, blustery climate, Chacabuco grew into the industrial merluza port as the other industrial fleets and shore plants followed suit. Another one hundred artisanal fishermen headed down to settle Aisen's inland sea to be nearer the market. But only twenty of the hardiest souls lasted in these foul weather conditions and treacherous currents.

Huge volume was necessary to make business profitable in Aisen, explaining why only industrial-sized companies had the economic flexibility to transfer there. Because of the distance and inaccessibility, the fish were destined for the frozen market. However, most of the fresh merluza buyers, who were smaller scale, knew artisanal production would become unprofitable if sold frozen. They could not shift to Aisen—a risky business with too-narrow profit margins, vulnerable to increases in operating costs.

Quotas hit during this precarious period of industry overhaul and amid recuperating stocks. But quotas were to hold for five years, keeping official landings down and allocating four times more fish to the industrial sector to the chagrin of the artisanal fishermen. The new quota system roused ire and brought all parties into furious debate. Even artisanal leaders from the two regions, X and XI, began quarreling between themselves as they were awarded unequal portions of the far smaller artisanal slice.

Many pointed out a quota system had failed dismally years earlier for the beloved national favorite, *loco*—an abalone-type shellfish—leading to rampant black markets with still no real control to this day. But managers blithely maintained their tight merluza quotas, continuing to focus on a collapse blamed on overfishing, and pointing out the diminished landing size of each individual merluza as proof. Both the industrial and artisanal sectors were collectively shaking their heads in exasperation. Their groups unified somewhat to fight the government for higher quotas. Artisanal fishing leaders began lobbying alongside industry representatives, as they simultaneously argued that merluza had come back.

This artisanal/industrial alliance is shaky. Their common ground is based on merluza having recovered at the moment. They are contesting stringent government merluza quotas that limit fishing to just a few days each month.

The industry's deep pockets have funded three artisanal leaders to lobby the sub-secretary of fisheries. So, in a prevalent global pattern where artisanal leaders act as liaisons to industry or conservation, they have been accused by their own folk of working all sides. Merluza fishermen feel that the industry has lined the pockets of their leaders, paying them off to ensure that only shared interests could be aired in front of the sub-secretary. Despite these fears and accusations, the three leaders did manage to address the sharp imbalance in the quota and attempt to wrest some away from the industry to gain an equal half. They pointed out to the sub-secretary that before quotas, 70 percent of the total catch—based on tonnage—had been artisanal. The sub-secretary still refuses to budge; an enormous catch of biomass being exactly what worries the biologists.

Islas Galas Life Continues

Back in Galas, seemingly detached from all the political complication, day-to-day life continues, and I'm getting to experience the realities of frontier lifestyle—the backbone of the business in this raw, wet region.

I'm visiting the only house in Galas with a shower, something I eye enviously but have no opportunity to try out. It is the family home of David who works part-time in radar for the Chilean Navy. His technology background led to his being trained to operate a government computer system meant to test the locally abundant shellfish for deadly red-tide dinoflagellates. He cannot demonstrate the computer to me because it's perpetually in Santiago being repaired. Instead, David is playing me five of the various Chilean television videos about the bleakness of existence in Galas. The titles, such as *Weeping in the Rain*, reflect the themes of hardship and woe. David sits on the edge of the couch with one child standing between his legs watching her father move across the screen as the star of this segment. In one scene, David points out Don Humberto, who is Galas's official artisanal leader, their current liaison to the fishing industry.

Several of David's other children are watching too. Drawing his daughter up on his lap, he says, "I'd promised my wife, Tina, when she arrived, a house by Christmas, so I began saving all my money. We lived off what Tina made at the school where she was caretaker of the boarder kids from the other islands. Many nights she slept in the dormitory with them, and baked daily school bread." David himself constructed this house, complete with that shower. Upon seeing it, I automatically fantasized hot running water. An illusion, it is really more of a cold interior rain.

He also demonstrates how to use his chemical toilet. They are now dotted throughout the community, donated by one development project. The catch is that chemicals to run them were not included and must be imported from civilization via the ferry, a cost. Besides being free, the chemical toilets have been adopted because the fairly pristine waters around their islands are where everyone fishes, washes their gear, and plays. This system is far cleaner and less pungent than the simple raised outhouse emptying onto the ground still used by some families. Locals are investing in this green technology, although I don't know where they dump the transformed chemicals, or how long they will prioritize buying chemicals in tight times.

At Gloria and Ladiseo's house, there is a separate little tool shed off to one side where the chemical toilet is ensconced, and it's biting out there. This is my excuse to my tropical self. After several days, in the warmest part of the afternoon, out of hospitality and concern Gloria says, "You have to get clean!" So she boils up a big pot of water and Ladiseo balances it down the steep steps and out to the shed where he sets the steaming pot by a bucket half full of cold water to mix in, alongside outboard motor, longline gear, and toilet.

"OK, Sarah," I'm thinking, "You're grimy." Chilling, in a rush I strip myself and crouch under a thin stream of warm water I'm pouring from a cup, conserving the supply, adding just a dash of liquid soap. Once rinsed and back into fresh clothes, I feel great, fresh. Cold is not my camp style; I've lived for months on end bathing in South Pacific tropical seas. I had to recognize that in David's eyes, running water, no matter the temperature, was a big deal. I laugh to myself, remembering how I came to understand running water as a big deal during fieldwork in 1983 Madrid—pre–European Union development. A sign hanging off a balcony in central historic Madrid read *aqua corriente*—running water—and the place became our hostel; at first we nicknamed the owner Mama Corriente before becoming grateful.

Other people could have created a catchment tank like David's for his sink and shower. Given plentiful rain, this system would work well. But then I realize that David is an innovator among migrants, who tend to be more daring personalities, more willing to morph than those who stay put. Ladiseo says he and Gloria are probably going to build a catchment soon. David had been the first in Galas to construct a house from milled lumber and with glass windows. Ladiseo and Gloria quickly followed that example, the result being this cozy home with picture windows facing the harbor, golden wood walls unpainted and fragrant. The old original huts of piecemeal fabrication—bits of plastic and driftwood often ingeniously fitting into the construction puzzle—reflect the initial harsher camping lifestyle of the first settlement years of the boom.

All ferries come right inside Galas's sheltered anchorage, an area cupped by three concave islands. The islands are stunning, their high rocky promontories, wooded, dramatically jut up from the sea forming this deep circular harbor. Ladiseo and Gloria sited their house at exactly the innermost point, looking across the whole sweep and northward. Gloria and I are having tea and she's telling me part of her compelling life history. She takes me into their bedroom to admire the vista and says, "I'm going to do a needlepoint of this great view. Take a photograph for me to work from, won't you?" I oblige. She is my image of a complete frontier water woman, from strong diver, to home builder, to needlepoint artisan, to baker, to shopkeeper, and even occasional bartender downstairs.

People live all along Galas shores, but most have settled either around Toto where Gloria and Ladiseo are located, or on the neighboring island, connected by a boardwalk, where David lives with the school, church, and recreation center. To get around, you literally wave and shout to a passing boat to hitch a ride. To go back home, I try large arm gestures at a boat zipping by, and sure enough it heads to shore. A guy jumps out and pulls the boat ashore, "Climb on it!" It's neighborly and reciprocal, and free, since rides are too infrequent to support a water-taxi service. None of these open fishing boats are over the artisanal maximum of eighteen meters. They're the community vehicles or workhorses. Neighbors help offload each other's boats. Many extended families run housefront stores, more by way of purchasing wholesale goods in bulk to cover transportation costs of living far from the source of consumer goods.

The Navimag ferry sailing the length of Patagonia from Puerto Montt to Punta Arenas pulls in here weekly and lowers its nose to let off people returning with their goods. But no vehicles can leave the hull's belly at this roadless stop. The smaller Quellon ferry, departing from the foot of Chiloe, also disgorges locals with goods in their arms and on their backs (figure 3.3). Occasionally, there's a traveler, like me.

When Navimag arrives, Gloria gets help unloading supplies she ordered to stock the little store and bar she started on the ground floor. She's justly proud of the big investment in the bar's wide-screen television complete with video player. They crank up a generator a few hours each evening to power lights and the show. I'm sleeping on the floor mattress in a room kindly vacated by one of her brothers to make private space for me. Each night, I immediately drop into sleep through the distant rumbles of an urban thriller sound track rising from below.

Figure 3.3. Quellon ferry stops at Gala s with time to buy snacks. Courtesy of the author.

Up steep steps after organizing new beer and soda to fill the bar, Gloria and I reach the family space, the architecture accommodating the sharp slope of the land. On this level are three bedrooms and the kitchen. The concept of flies as bearers of disease in the kitchen has not been instilled in Gloria and the others. I'm not accustomed to the large size and inevitable density of the flies during Patagonia's brief summer months, although I love village life. Trying to be casual, I point out a juicy, buzzing one who's hiking in the unpeeled tomato salad. Gloria, blasé, forks him onto the floor, where he struggles, and takes off again. You can't be a pioneer without pioneer sensibilities, I'm thinking. None of her leftovers are covered to protect them from flies, or cupboarded to keep them from rodents. Government posters warn that hanta virus might arrive on shipboard rodents—a pretty deadly one spread to humans by rodents nesting in plant fiber roofing throughout the Andean region—but so what? The tomatoes are full of flavor. I eat heartily.

Galas households like Gloria's depend on a neighborly sense of shared labor and common survival. Just as Galas islanders hail rides and help each other in the constant loading and unloading of boats, they hold something of the feeling found among people thrown together after a disaster (figure 3.4). There is the emotional tone of making do and being able to be self-sufficient if everyone shares skills and energy. Single mothers make it here, as do old single men. Of course there's gossip, jealousy, competition, and cliques, as in all villages. But the need for making do and creating a better life override many of these feelings—at least while people are still few and share institutions, for example, by having one church and one school.

The father of this church, Padre Antonio Ronchi, was the community glue, counseling for love and friendship, holding together dreams of a decent livelihood, finding funds for development projects. He was the spiritual heart of the community. The vacuum after his recent death left behind a community diversifying far beyond its roots as the pioneer merluza fishing camp it was when he first arrived.

The Return of the Fish

By the end of the merluza crash, more than one hundred small plants buying fresh merluza closed in Puerto Montt. But within three years, these buyers were back. Catch sizes per boat had recuperated in the southern inland sea canals around Galas. With the happy return of the fish, people feel that the merluza business is here to stay. With renewed vigor they scramble as stakes have increased along with the number of stakeholders with competing interests.

"Nobody believes the other," says a Region X fisheries official I've known for a while. He's busy trying to broker discussions by telephone from his Puerto Montt office high up on the hill overlooking the expansive bay.

Figure 3.4. Fishermen preparing to go out for merluza from Galas. Courtesy of the author.

Oppositional groups include artisanal versus industrial; artisanal from Region X versus Region XI; small fresh merluza buyers versus industrial plant buyers; artisanal and industrial sectors versus the sub-secretary of fisheries. Alliances and competitions are impermanent, continually forming and dissolving in not-so-clear-cut ways. My mantra resonates: the groupings shift in response to economic, political, and environmental changes.

When the industry entered the fresh *merluza* market, for example, six fresh merluza buyers from Region X formed their own export group. By 1996, ten companies had been motivated to join. They hoped to gain political clout when lobbying the government for higher artisanal quotas. One of the members—SALMAR—set up operations in Chacabuco exclusively to export fresh artisanal merluza during the warmer half of the year from October to April, closing down during the stormy winter when artisanal fishermen can catch little.

Of course the industry based in Aisen wants a merluza monopoly, and thus wants SALMAR and all buyers who deal in the high-quality artisanal fresh fish out of the market. The industrial trawler-caught fish, which often sit fifteen days at sea in a cold-store hole, are much diminished, lacking texture. They must be filleted, washed, and decapitated to hide the pale, bloodless gills and clouded eyes—telltale signs of an old dead fish—to be sold at a far lower price in Spain's huge and famous fish market, Merca Madrid. Year round, the Spanish giants with their freezer plants in Chacabuco and direct markets at home are avidly buying the artisanal catch as well as the catch from their own fleets (figure 3.5).

Adding to the fresh/frozen rivalry is a new international wrinkle for Chilean artisanal fishermen beyond their control or ability to protest: unstable prices in the fresh merluza market. Madrid began 1996 with an oversupply of fresh merluza, so the price dropped. The glut arose when Namibia decreased their frozen quota while increasing their fresh quota to target the more lucrative Spanish market.

Artisanal leaders still feel cheated on price by the industrial buyers. A leader from Galas meets me in secret to whisper, "Our buyers tend to pay relatively little. Once in Aisen, the industry pays US$1 per kilogram, only one-fifth the price offered in Madrid. Strong catches mask this problem, while weak catches highlight it. Then industry, with their unfair group pricing power, deflects our protests to government."

By trying to organize new protests in conjunction with artisanal leaders, the industry hopes to refocus the artisanal lobbying efforts away from pricing to government quotas. Thirty artisanal fishermen from Puerto Montt's most politically active group—Anahuac—stage a hunger strike in 1996 against the stringency of the whole merluza quota faced by both sectors.

Artisanal leaders also point to the industrial fishing during merluza's three- to four-month reproductive season, when they aggregate at sea along the coast and are easiest to catch. They construe that government and this big business are in alliance so that government management focuses on the artisanal juvenile catch in the inland sea rather than industrial fishing during the merluza spawning aggregations—a universal no-no from the perspective of fisheries management.

Through all this, the government remains unyielding on quotas, still claiming overfishing, focusing on the lack of juveniles, and heavily reining the artisanal sector. The sub-secretary of fisheries—who ultimately decides Chile's quotas—has refused to significantly raise the overall quota, insisting that merluza stocks still are in a reduced state.

Sea Lions with the Juvenile Merluza

Region XI's monthly merluza quota is opening soon. Four buyers have arrived in Galas and are anchored in the inner harbor, near Ladiseo and Gloria's house. The holds of their broad-beamed thirty-five- to forty-five-foot vessels are empty, ready to receive merluza for the run to Chacabuco. People drop by their favorite buyers to exchange greetings and arrange business. Marta with her daughter comes to visit one buyer while I'm talking to him. Marta says, "Who we decide to sell to is based on friendship, bringing us

Figure 3.5. Chacabuco's industrial merluza base for processing. Courtesy of the author.

goods and advancing money. It's also trust in weighing the catch because prices don't vary much between buyers."

Excitement is building. Many fishermen, like Marta's husband and brother, have already headed out "to test the waters"—namely locate the hottest merluza grounds so as not to waste a precious moment at the start. Legally, this is a gray area; harvesting prematurely isn't. There's always a fear here that regulators might show up at Galas to observe.

"If they come home with catch between the test and the open, we keep it hidden in coolers," says Marta.

Ladiseo later points out several such coolers, so I learn to recognize the system. Buyers, meanwhile, stay innocently empty until the radio announces: "The quota has been met for this month." With that official close—true to form in just a few days—I watch everyone plying back into Galas loaded with merluza. Substantial numbers of sea lions also gather in port around the commotion as boats take turns coming alongside, fishermen handing up the catch, fish by fish. Each buyer eyeballs a smaller fish to judge if it meets the legal size limit (60 centimeters minimum) and some are held against a measuring stick. Buyers throw the many undersized merluza overboard to the barking delight of the sea lions, who toss their free meal back and forth like a ball, playing and enjoying the feast.

Fishing Regulations and Marketing

Based on Chacabuco landings in 1996, the government biologists estimated that over half of the artisanal catch was juvenile. The industry was playing up bad juvenile overfishing; the artisanal fishermen feared their share of the quota was about to be slashed.

After two years of saber rattling, the industry and biologists start to be impressed by the official artisanal landing record. It shows that the landing of juveniles had practically stopped and the average merluza had shot up in size. The sub-secretary of fisheries also interprets these landing statistics as the reality in the water. In truth, the artisanal fishermen and their buyers have been treating juveniles as by-catch all along, throwing them overboard dead to the glee of the sea lions. If the first sale—buying directly from fishermen—rejects tiny juveniles and pays a slim amount for the borderline legal, then landing figures will reflect that market. Legally, only 15 percent of the merluza brought to the processing plants can be juvenile, since there is a parallel need to meet European size restrictions.

Fishermen here were responding to market changes, not policy, and many juveniles were dying off the record. Thus government studies, which rely on landing figures in Chacabuco, easily misjudge actual numbers of juveniles. They completely miss the real picture of what is swimming in the water; what is caught at the fishing grounds where the obviously undersized are abandoned; and what is offloaded at a fishing port like Galas but ends up in the bellies of sea lions. Markets universally sway the decision of what to catch. Because of this, conservation groups—such as the NGO Traffic—take the angle that buyers should be the main ones managed instead of fishermen. Governments would do well to listen.

Complicit with the buyers, fishermen do not bother to adhere to the merluza quotas and steeply surpass them. The disregard is once again an expression of frustration, a common form of resistance among fishermen facing management from the top-down where they have not participated in policy design; they feel the regulation benefits neither them nor the target species. Throughout Chile and into Peru, artisanal trust in the fairness of government policy has dwindled because fishery management is notoriously colored by manifold direct business links between export industries and government. Again, some Puerto Montt area artisanal fishermen who are politically attuned and less focused on survival, literally march against the artisanal quotas, as well as the low prices being paid by the industry.

But because of shaky merluza export prices, and as a hedge against another merluza collapse, the industry and Puerto Montt's smaller buyers/exporters are expanding into other marine products, such as other types of finfish, mussels, sea urchins, and clams. The artisanal fishermen call the industry and exporters *pillo*—clever like a fox—and feel it's wise to emulate diversification. In Galas, like in the other caletas, Gloria and Ladiseo must dive and fish for a variety of species to supplement merluza.

The Motor of Islas Galas: Padre Antonio

I wanted to meet Padre Antonio Ronchi, a Catholic priest from Milan who settled in this region thirty years ago. He devoted himself to these outer islanders. People here still refer to Padre Antonio as their "motor." He raised grants, like the one for the US$30,000 computerized shellfish tester, and companion funds to train three people in the community, including David, to use it to check for red tide contamination in shellfish. Red tide is common in Galas; shellfish are abundant yet vulnerable to contamination. The padre hoped to ensure safe local consumption because infected shellfish are deadly to eat, yet exhibit no external warnings, appearing healthy. With this sophisticated piece of technology currently broken and away being repaired, Gloria pragmatically says, "We continue to eat shellfish anyway. They're available and delicious."

About a year ago, David and the two others who did the training made a pilgrimage to Puerto Montt hospital where Padre Antonio lay dying. Only sixty-seven years old, his heart had failed. Everyone still feels incredibly sad, missing his jolliness and hope.

"He went around singing and seeking friends. Besides the church, he was always campaigning for us, finding funds for everything from schools, to canneries," says Ladiseo. The strong and expansive spirit of the padre lives on; I see tears well up as locals talk about him. I just wish I could have been here a year sooner to get to know him.

"Now our motor is broken," says David, a bit demoralized. I ask him about Galas finishing community projects that Padre Antonio initiated, such as running electric lines from the giant generator he'd procured for them. "Yes, we'll do it, and this will be our way to thank him."

Padre Antonio had asked, "What is poverty?" The people have been labeled "poor" by the rest of the country, and they themselves need to shake off this stigmatism and self-doubt about their real worth. He'd worked in a participatory way among the little fishing colonies, traveling island to island asking what was needed and helping to get it. His aims were social justice and developing a local sense of pride in building up their communities, constructing schools and churches, bettering their livelihood. He gave pep talks and provided the start-up funds and material for community projects. And then, Padre Antonio would switch roles and head up to the cities to pitch the projects to the central church hierarchy. Educated and charismatic, he functioned as a liaison respected by both worlds.

Lots of people liked telling me how Padre Antonio would go out on little boats and sleep curled up inside. He wouldn't wash his habit for maybe a month at a time, so the wooly thing hung to his feet, odiferous, and he didn't care. He ate whatever people gave him in good humor, truly following the Franciscan life of poverty, helping others with compassionate generosity.

Padre lent everyone confidence, from the kids to the adults. Back when the camps were really rough, men drinking, rowdy prostitution, more of a runaway convict type of place, he'd moved in. The men were living in these tent or hut assemblages of canvas, cardboard, scraps of metal, plastic, and driftwood. Padre Antonio would pass from shelter to shelter, asking what people wanted, and when some of them said they desired God present in their lives, a church to attend, he helped them build it, but didn't stop asking.

CONCLUSIONS

Carlos Castaign of Puerto Montt notices a great match between his native merluza and the cavernous Spanish market. So in 1981, he started exporting the high-priced merluza to Spain. The fisheries would grow into a phenomenal boom, converting men into fishermen and spurring colonization of Patagonia's cold isolated inland seas.

As the Spanish market swallowed up the catch at a fat price, more and more small-scale buyers desiring direct access to merluza began their own companies. Buyers needed more fishermen to feed the market for the prized merluza. The market creates the fisheries.

Merluza is a great example of spontaneous, unregulated fisheries. It bloomed in the context of rampant privatization and retooling for export that characterized the 1980s Chilean economy. Chile's neoliberal economists were cultivating export markets—such as merluza—while facilitating foreign investment. Fervently, they believed in implementing policies to encourage export that would lift Chile out of an economic crisis that found millions seeking jobs.

Desperation provided the perfect timing to recruit for a merluza boom. Fishermen as well as jobless men were contracted to migrate into the rainy remote inland seas to target merluza. They were willing to endure a cold lonely existence for the sake of potentially great money, and by the height of the boom, huge numbers had joined in. Exact figures for the fishermen and the merluza they caught remain unknown. For one thing, this first-time boom was spontaneous and unregulated.

The extreme lag time before government introduces regulations for any new boom fishery is a common occurrence worldwide. Merluza took seven years to crash, ruining more than one hundred small buyer/processing plants in Puerto Montt, yet it took Chile a couple more years to codify the 1991 Fisheries Law aimed at artisanal fishermen. Management quotas for merluza began long after the crash, during the precarious period of industry overhaul. The industry, which survived, decided en masse to move their buying/processing operations deep south to Puerto Chacabuco. Ironically, this transition and the new quotas system coincided with the surprise return of merluza.

The restrictive quota system focused all parties in furious debate over how to share the merluza. Merluza's industrial and artisanal fisheries shakily unify to fight the government for higher overall quotas. Since catch sizes are recuperating, small buyers reappear in the little ports, many reselling to the industry now down in Chacabuco.

In a rivalrous turn, the industry points to the official artisanal landings in Chacabuco to prove that artisanal fishermen caused a serious dearth of juveniles. They lobby for government studies, which classically end up supporting the position of those who commissioned the studies. In reality, artisanal fishermen have been catching juveniles but are throwing them away as unmarketable, to the delight of sea lions. Sea lions circle around the buyers in the little ports who are tossing the borderline merluza overboard because the industrial market refuses to purchase and process under a certain size. Thus, fishermen are responding to what buyers are willing to purchase rather than top-down regulations or in-the-water reality. When this landing data is used against them to set artisanal quotas, they bridle at the sense of economic injustice and incongruent visual evidence of the merluza availability. They feel justified in ignoring or bending regulations. While quota fishing, some are stashing merluza away before the official openings, or ignoring the whistle to stop when the quotas are reached. Smuggling in merluza for the industry from zones out of turn is also fair game.

EPILOGUE: LEGACY OF THE MERLUZA BOOM-TO-BUST CYCLE

The creation of a large group of artisanal fishermen currently underemployed in Patagonia lingers as the boom's legacy. Galas and the other colonies have transformed themselves in the boom's wake into small family communities, complete with churches and schools. To get through they need income from shellfish diving and other fisheries in between merluza quota fishing.

Quotas remain in force, having endured countervailing political pressures. With the justification of protecting the juveniles, the industry has been able to sink the artisanal quota down to only one-quarter of their own. Artisanal fishermen meet their quota within the first few days of each opening. Then they lack work. The industry, on the other hand, often cannot complete their quota restricted to the ocean side. Artisanal fishermen, however, continue pushing to redistribute the quota between themselves and the industry to match historic and current fishing effort.

In the inland sea of Regions X and XI, no enforcement limits the number of merluza fishermen, despite the merluza registration lists being full and closed. Outsiders continue to migrate down, moving into the communities and illegally joining the merluza fishery.

With artisanal fishermen's future contested and unclear in the inland sea, a few intrepid and desperate fishermen have gone deeper down to Punta Arenas, Region XII, near Cape Horn, where the merluza registration list isn't full. Who would wish to work in the face of constant gales? They have registered their open boats for the open water's horizontal howling winds, and frigid tempestuous seas, which place insurmountable limits on time spent fishing. Closest to Antarctica, Region XII's climate is Chile's most extreme at sea, perhaps allowing the boldest fishermen five days a month of fishable weather. Region XII's tiny fishing fleet never meets their merluza quota, although they have been allotted the smallest share of fish for all three regions.

One of the rare Chilean women in top-level fisheries management, Isabela reigns over Region XII's fisheries as manager. She's an elegant, assured, and articulate person. I'm excited to meet such a woman when she grants me an interview in Punta Arenas. Large charts cover the vast walls of her department inside one of the town's colonnaded colonial buildings. We compare life with Chile's other extremity, Iquique, that also has been awarded a duty free zone, including cheap vehicles for residents, in an attempt to encourage settlement. Weather cannot help but be a topic of conversation. Average weather for Punta Arenas fishermen only affects fishermen in Chile's more northern calmer regions during the stormy weather patterns of a major El Niño.

"My mission is to protect Chile's precious parent stock of merluza, especially because of the intense fishing of juveniles in Regions X and XI. Luckily, nature is managing the stocks through wild weather," says Isabela, assured she can safeguard the merluza and that they can grow much bigger and reproduce in peace.

Isabela notes how overfishing merluza in the inland sea has been considered the official cause of the merluza crash. Managers frequently blame fishermen for a crash, being the most apparent cause. Yet El Niño's cumulative effects may have been a root cause, this natural phenomenon playing powerful and still unclear ecological roles among marine life worldwide.

Isabela recommends a pilgrimage to a big colony of Magellanic penguins along the gale-swept scrub coast so I can get some sense of the weather conditions all beings face at the end of South America. The next day I head out to their nature reserve. The large colony is cordoned off with trails that skirt their burrows a few hundred yards inland. The winds are howling, and the penguins are braying (they are thus nicknamed "jackass penguins"), as I watch them digging out their dirt burrows and hunkering down in families sheltered from the piercing gusts. Babies take turns daring to stick their little black-and-white heads out for a few seconds. Parents returning from fishing forays make flying leaps out of the waves, braying into the wind. Forming droves, they time their dash back up the rocks and across the colony so they can shoot directly down into their home burrows.

Back in town by evening, hands curled around a bowl of fish soup, it's impressive contemplating everyone braving these seas. The rewards must be great. It makes life on the inland sea feel calm.

LETTER: HIGH ABOVE RIO ELQUI TO THE STARS AND PETROGLYPHS OF COCHIGUAZ

Dear Bruce,

After a quick visit from Patagonia back up in the northern desert port of Iquique, I catch a flight heading to Coquimbo. For only the second time in my life, pilots let me ride up in the cockpit to photograph their tremendous panorama. Desert mountains green into the coastal hills of Norte Chico—Region IV being called the Little North. I take nonstop digital pix of raw earth's shifting colors and shapes, as well as a few of the pilots, who are teasing each other about becoming stars. Amazingly, they let me stay up front with only a lap belt, breathless for the landing up into Valle de Elqui. I've never had the opportunity to land in a jet cockpit before, with the full force and speed of approaching tarmac, the runway lines converging. It's an exhilarating way to begin at a field site. For the finale, I made a Cargo Art shot of the control panel buttons looking back at me like eyes.

I've been making my Cargo Art spirits via photos, lacking a studio. I've also been collecting terrific anthropomorphic tools from the colonial period and abandoned bits in the desert near nitrate mines. Keeping me company so far are a blue enameled teapot dented and rusted into a personality, a tiny British iron that opens partly for charcoal, a wooden-framed saddle stretched in leather now desiccated with the metal arched front appearing like a New Guinea mask's protruding lips when I stand it upright, an old shoe made with wooden nails whose tongue hangs out, and a miner's Pata Wells oil lamp with a suggestive spout with that surprised "ohhhh" look. With the addition of Solomons green sea snail operculum eyes, I already know I'll turn him into "Never Schooled in Romance."

So I've settled into life in Coquimbo with my Cargo creatures, lodging in university guest housing up a steep hillside overlooking the entire industrial port. Hiking farther up to the top along goat tracks, I reach a famous pirate lookout and enter their cave. Imagine holing up here with parrots by bonfires. By following paths cutting down to the water, I can commute to the Universidad Católica del Norte where I'm working with the Department of Ecology and their leader, Dr. Wolfgang Stotz. He's a wonderful, red-bearded man of German descent, considered a top expert on Chile's upcoming system of fisheries Management Areas. We're investigating how local fishermen relate to Niño and Niña impacts on their nascent Management Areas.

Coquimbo makes a great field site because it's where the Rio Elqui dramatically flooded during the recent El Niño, burying all bottom life in a thick mantle. Sediments choked out the entirety of clam banks where the fishing co-ops had been carefully managing their harvest. Then when the colder-than-usual waters of La Niña struck, the recuperating stock of clam babies were rapidly picked up by poor urban folk illegally. Now in a common refrain, divers say, "They've left few behind to grow." Tension is high as a result.

It's beginning to be springtime in this exceptionally rainy El Niño year. Outside my bedroom window a tree blooms. The scrub desert here is extraordinary, perhaps once a decade bursting into this flowering spectacle. Again, I can't resist constant photography. The goats, and the wives of fishermen who herd them and make fresh white cheese from their rich milk, are contented. I love the cheese.

Saturday, I board a bus heading up past the new dam Puclaro. It is being built post-Niño floods, with the intention to irrigate the grapes and miniature papaya plantations. The dam will also prevent any future flooding into Coquimbo Bay. With the first snowmelt, a lake is beginning to fill behind its dike. Continuing higher into the mountains, we pass the side route to Chile's major astronomy observatory of Cerro Tololo where at 2,200 meters in the driest crystal air, huge telescopes are poised at the heavens, one with a lens four meters in diameter.

We keep climbing through seas of grapes for *pisco* winemaking. I jump down by the dirt road leading off into Cochiguaz Valley. A local bar owner doubles as *taxista* and up, up we go. At 3,000 meters, this place is called *mítico*—mythical—a magical place for ancient petroglyphs and stargazing. I've timed my high visit because tonight is to be a major Leonid star shower. You would love this pilgrimage with your knowledge of astronomy. There's no electric light pollution, only vast dry skies of clarity.

Cochiguaz is considered an energy center by Chileans, like Sedona in Arizona. An international collection of New Age older hippies, and ex-Santiago city folk have settled in and are practicing healing arts from aromatherapy to Reiki.

At the far end of Cochiguaz, in this lucidity of light—the complete opposite of Santiago's foul air—lovely Roberta in braids brews herbal perfumes and chants to the stars.

I ensconce myself at her hostel, El Alma Zen—the Zen soul, *almacén* meaning store—where Hindu multiarmed gods painted on the walls rest by words declaring all in the name of Jesus. Roberta is the wife of the owner and tells me how her husband has escaped being a nightclub striptease kingpin with a famous Santiago mobster as partner. Doctors declared that his blood pressure was about to kill him, so he was "allowed" to retire, not easy to do from any mob in the world.

Toward dinnertime, I'm drawn a short walk uphill by the smell of baking bread. There's Gabriela, sliding loaves out of her outside clay oven by an adobe house. It's old-style adobe, five rooms side-by-side attached in a line, sharing one long porch, each room with a door facing the yard. A couple of men are rocking under the eaves. Gabriela invites me over, saying, "This grinding stone is what adds the special flavor." I can see it's probably pre-Colombian. I enjoy the warm bread as we talk about life in this valley. There's a whole-wheat baker at a commune nearby, and the whole scene is mellow.

At 2 AM when the star shower is supposed to start, swathed in Blue Cloud, I climb up into El Alma Zen's Olympic-sized amphitheater (soon to be a swimming pool) and cuddle back. Alone, I feel pulled up into the Milky Way, riding multiple streaking stars a minute for over an hour. As the morning sun rises over the mountain shadowing El Alma Zen, I hike to a massive petroglyph-covered boulder set in a horse field. It's threaded with snowmelt rivulets. A gaggle of geese are ducking along, cracking through thin ice of a boggy springtime. As sunlight stretches forward and covers the boulder's face, the carved guanacos jump into dimension.

It's time for me to head back to the Caleta Coquimbo's Sunday fish market. I start out on foot, soon hitching a ride back to the sea with a Santiago family—the father discussing his homeopathy practice. By noon, I'm in another world, strolling around tables heaped in raw wet seafood. Families are relishing plastic cups full of shellfish—minus those clams killed by El Niño. Sea lions and pelicans gather around fishermen cleaning the catch to beg their share of fish guts. The debate in Santiago fisheries circles might lead to satellite positioning systems being installed on industrial vessels. At least this could keep the industrial boats outside the caletas' first mile.

A pair of sea otters have moved into Coquimbo's caleta, swimming on their backs among the rafts of pelicans, so I'm going to watch them before returning to my hillside. With a few more days of interviews, I'll be heading to Paracas, Peru, for the month of September.

Hoping you, Mango dawg, and Kipling Kat are fine at Fanalei.

Be well, happy composing, my love,

Saludos from the Little North,

Sarah

Chapter Four

The Dance of Macha Divers and Waders around El Niño's Flood

DRUMMING UP THE SANDS FOR MACHA

Facing into the current, diver Danilo taps the bay bottom with his hands like a drum. Sand flies up and is swept away. In this way, Danilo is excavating a shallow trench to expose the valuable bivalve, *macha*, along with other uninteresting heavy objects left behind. I'm sitting up on the surface in a bobbing boat with the compressor chugging away as Julio, the onboard crew, feeds Danilo lengths of compressor hose as he bellies along the bottom.

"Danilo's down about ten meters now. He's got to negotiate the surge and powerful currents ripping along the bank. That's why he strapped on so much weight, so he can hug the bottom." Julio points to bubbles rising to the surface and continues, "Follow Danilo's air. You can watch him crawling forward, digging in that longline to pick up macha."

After another forty minutes, Danilo is back onboard stashing their macha, shivering a little in his wetsuit. He turns to me, "It's not bad. There's light at this depth, and I don't have to be worried about pressure. Those are concerns for divers hunting shellfish in the deep."

EL NIÑO TORRENTS AND EMERGENCE OF GLOBAL THEMES

During the relentless El Niño rains of late 1997 to February 1998, havoc breaks loose. The normally narrow and flat Rio Elqui flowing out of the Andes into Coquimbo Bay swells and bursts over its banks, funneling decades of accumulated debris down into the bay. Producers of famous wines and spirits such as the Pisco Elqui, have been planting vineyards high up into the Elqui Valley and along gorges right down to the water's edge without restrictions. Nobody had considered the possible sweeping force of El Niño torrents, as it has never happened in memory.

Stunning everyone, El Niño's surprise flash flood of Rio Elqui, has wiped clean the valuable macha—a clam-type bivalve. Thick blankets of sediment settling on top of the macha have choked them to death in their sandy beds. In the aftermath, Coquimbo municipal workers labor daily to clear the bay's beaches, lugging away old wine cartons and plastics, dead animals, and heavy branches. Tourists—especially the main draw crossing over from Argentina—stay away from the stinking mess for at least a season. Ravaging hotels and other tourism interests, it's really Coquimbo's urban artisanal macha divers and waders who bear the long-term repercussions.

After El Niño's flash flooding smothered these banks, I went back and found some of the divers, like Danilo. Almost none of them want to switch dive techniques to be able to target the deeper types of shellfish. This makes sense for anyone accustomed to light and shallows, without the risk of being bent—decompression sickness. Instead, they rely on their own strength in navigating the bay bottom beyond the surf. But both divers and waders must migrate until Coquimbo Bay macha can recuperate. This underscores an important survival strategy of coastal people who shift locations when an area needs to recuperate.

Dividing Macha before the Flood

Divers, such as Danilo, are the more recent type of macha harvester—called *macheros*. Divers are based in two of the bay's three caletas, Caleta Coquimbo and Caleta Peñuelas (figure 4.1). Caleta Coquimbo, the largest, is an urban caleta poised on the south side of the bay. It's comprised of various fishing arts whose fleets are manned by many migrant fishermen who have settled down here. Unlike the small Peñuelas beach community of interrelated Peñuelino fishing families, members of Caleta Coquimbo live among hilly urban slopes of Coquimbo's roughest neighborhood.

Waders have the deepest historical roots. For over a century, Caleta San Pedro on the bay's northern shore has been a community of waders subsisting on macha in the shallow intertidal zone. It's out by the dunes on the outskirts of La Serena. Serena, the second oldest Chilean city is the charming capital of Coquimbo Region.

Waders, both men and women, walk out from shore at low tide, thus they are called *orilleros*—*orilla* meaning shore. The men usually wear wetsuit tops and masks, working up to their necks in the deeper areas along the wave break. Tapping with bare feet to excavate, the men feel any sharp hard lumps that might be macha. Meanwhile the women, still in street wear, stay in

Figure 4.1. Coquimbo Bay and Rio Elqui map. Adapted from base map from U.S. National Imagery and Mapping Agency, Perry-Castañeda Library Map Collection, University of Texas, Austin.

thigh-deep water doing the famous twist of the orillera—a rapid hip-wiggle dance that burrows their sensitive feet deep in the sands to uncover macha.

Part-timer waders also join the ranks, entering the water in regular clothes with no further investment to protect themselves from the cold. Everyone uses the long, narrow mesh bag called a *chinguillo* for carrying the macha. Nobody needs other tools. Because the means of production are so simple, anyone with a constitution to withstand cold—or perhaps just anyone who is desperate enough—can try their luck. When fishing or economic times are tough, officials follow the cultural pattern of looking the other way as fishermen and the hungry ignore management regulations in order to survive. Consequently, jobless and other urban poor unaffiliated with caletas freely wade out to collect macha and seaweed disregarding any caleta self-management efforts or marine claims.

Unlike the divers' product, macha collected by the waders does not go to factories. Small-scale buyers who sell on the national market purchase macha direct from the waders. They freeze the macha in 250-gram bags for supermarkets, or sell them on ice in a typical mixed selection of shellfish for popular seafood soup.

Tricking Buyers

Pepe is a typical buyer for the waders. He's down on the beach with his small truck, negotiating among people whose hair is still dripping wet as they're emptying their mesh bags into mounds of macha on the hard, damp sand. As they make mutual agreements, the tide begins creeping back up the beach from its low point when waders were in full swing earning cash for the day.

Once Pepe has loaded up today's harvest and mentally calculated his profits, he tells me about an old waders' trick of putting smaller macha into freshwater to make their tongues temporarily swell up to appear larger." But when cooked, the macha shrink back to size," he laughs. "These waders don't dare do it to me, because I often buy on the beach and we know each other. But the tongue trick fools the public at the caleta market buying Sunday dinner. They're hoping for wholesale, and end up spending more than at the supermarket."

The lower the prices and more local the consumption, the more subsistence-oriented the fishery. You also see this pattern among Galician shellfish-wading women before Spain joined the EU and gained access to lucrative markets (Meltzoff 1995). In Coquimbo, the beach price for locally consumed macha was so paltry that there was little incentive to develop other methods for collecting beyond subsistence wading. The advent of macha diving took many decades to come to the area and was brought by outsiders.

In 1977, a diver fleet fifty boats strong migrated up from San Antonio in central Chile, Region V, in search of intact macha banks after having decimated their own. In a strategic pattern of survival management, fishermen have been migrating up and down Chile's elongated coastline leaving exhausted areas behind to recuperate, similar to swidden agriculture. This is equally true in Peru (Meltzoff et al. 2005).

Once inside Coquimbo Bay, the newcomers discovered only the San Pedro waders busily harvesting the shallows. They met zero competition for diving out beyond the waders and the waves. Underwater they found copious macha. The divers settled in, basing themselves in Caleta Peñuelas, boats kept pulled up on the beach at the central heart of the bay. Peñuelas' small, tightly knit fishing community of ten boats had been focusing on netting hake inside the bay. Peñuelas fishermen began learning to dive macha from their guests. After a handful of years, the San Antonio banks had recuperated, so most of the San Antonio divers drifted home, leaving behind depleted macha beds but a crop of new divers.

Although the surf seems to physically divide the grounds of divers and waders, it's a continuation of one big macha stock extending out under the waves. Thus, intense diving had damaged the amount of macha within the waders' range. To compensate for lost income while the macha recovered, government presented Caleta San Pedro's waders with a carnation-growing project. Whenever a project is delivered top-down to locals who remain uninterested in maintaining it, people tend to accept the funds as long as they last, then let the project fold by attrition. Here too, as start-up funding dried up, the flower business faded away, and wading resumed even though the macha were not fully recuperated.

Meanwhile, a San Antonio buyer soon arrived in Peñuelas to recruit divers to come join their erstwhile friends in San Antonio. Of course the far smaller Peñuelas group eagerly expected reciprocal hospitality. To learn what happened to those who migrated to San Antonio, I located Jaime and a couple other Peñuelas divers who'd set out with high hopes of finding their friendship network. They're on the beach rolling their boats out of the water on logs. Once the boats are put to bed, we head for one of the simple beachfront fish restaurants, eager to talk about what transpired.

Jaime shakes his head, "It turned tough quick. There were so of many of them diving. Through their eyes, we evolved into the competition, extra hands pulling macha out of their home banks. We caught hostile looks, ran into cold shoulders, and were socially forced out. But back home in Coquimbo Bay, nature had flipped in our favor. The monster 1982–1983 Niño had blessed our macha beds. Macha rallied and nobody in Coquimbo was diving them."

With the climatic variation in their favor, when Peñuelas divers returned home from San Antonio, they found beds brilliantly recovered. El Niño's warmer waters had triggered great spawning and driven off most predators unable to tolerate the heat.

With another external stroke of force outside fishermen's control, the globalized export market for macha began in 1986–1987. This a fine example of outside change critically interacting with local circumstances, such as the price of fuel, or the development of transportation and more advanced technology. The macha export market initiated a far higher demand and prices. Local processing plants began canning divers' catches for export; they previously had been consumed unprocessed nationally. The market rapidly encouraged expansion. After spending time alongside San Antonio's skilled divers, Peñuelas divers were seasoned and ready to respond. Their fleet blossomed, doubling to fifteen boats.

Caleta Coquimbo—poised on the south side of the bay—also embarked into macha diving. In contrast to Peñuelas, Caleta Coquimbo is the largest of the bay's three caletas (see figure 4.1). It's an urban caleta comprised of various fishing arts whose fleets are manned by many immigrant fishermen who have settled down here. Unlike the small Peñuelas beach community of interrelated Peñuelino fishing families, members of Caleta Coquimbo live among the hilly urban slopes of Coquimbo's roughest neighborhood. The caleta's newest faction, the macha diver fleet, drew in 70 percent new migrants, especially from the surrounding Regions V and III. They launched a dozen specialized boats onto the bay.

The combined dive fleets taxed the beds to supply the canneries. Unfortunately, a mild 1987 El Niño event brought a lot of rain to Coquimbo Bay but neither revived the macha beds like the 1983–1984 El Niño nor caused flash flooding to bury them alive as was destined to happen in the 1997–1998 El Niño. Instead, during this minimal 1987–1988 event, buyers busily contracted groups of divers who couldn't find enough macha at home to travel down to Patagonian waters for a lucrative loco boom. Migration in pursuit of the best markets is a basic Latino Pacific survival pattern. En masse, Coquimbo Bay divers again shifted survival tactics. One buyer arrived in Peñuelas and was able to recruit thirty boats; another buyer rounded up ten boats with willing crew from Caleta Coquimbo. Dive pressure was off the Coquimbo macha while they were away.

Jose Cariaga Leads San Pedro Waders

Around this time in 1988, Jose Cariaga—an outsider with social idealism bred during Allende, like so many nascent caleta leaders—arrived in San Antonio from the sprawling urban poverty of Santiago. He began by following local waders and copying their techniques. Like Coquimbo Bay, San Antonio's shore topography is ideal for waders. Both shorelines slope gradually to beyond the surf break.

Cariaga describes how he reached his present position leading San Pedro waders, saying, "As a youth, I was raised in state homes. I had nothing given to me. Once I got married, I felt pressure to earn money for my new family. That's why my wife and I left the city and moved just up the coast to San Antonio's prodigious fishing port. The poor know we can always wade to survive, no need for gear or diplomas. Wading is one of those fishing arts that just needs personal strength.

"We got off the bus and walked over to the wader caleta with our tent. They were kind to us, sharing food and camping tips with a naive pair of city newlyweds. Seventy of them had organized their caleta themselves. I happily followed them into the cold water. They'd lent me the necessary weights and mask, and were generous in teaching me the tapping dance, their art.

"But, it turned out we'd arrived at the beginning of the end of San Antonio macha beds, which were emptying out fast. We witnessed the dilemma of macha being grabbed by us waders and the divers, who formed an even bigger group. Their dive caleta had a fleet about thirty boats strong and the organization had more clout. Every day we helplessly watched their boats ducking in closer and closer through the surf bordering our wading zone. What could we weaker waders say? Nothing and to nobody!"

Cariaga continues with his wader saga, saying, "With meager macha left in San Antonio, in the summer of December 1988, I took it upon myself to fund my own exploratory trip for new beds. I didn't wait complacently for a buyer to come along and find us waders. Instead, I boarded a coach bound up the arid coast to Antofagasta, Region II. It was my first trip up so far north, and what a sea of mineral rocks and sands.

"Although I'd only practiced wading for about a year, I identified with it. For four days, I went around asking Antofagasta Caleta people about possible wading grounds. In sum, their advice was to backtrack part way to check out Coquimbo Bay, where they knew there was a macha market," he says.

Cariaga's next bus lands him at La Serena's lighthouse, demarcating the midpoint of Coquimbo Bay. The heart of La Serena is at the top of the wide, tree-lined boulevard running perpendicular up from the beach. Elegant Serena is coupled with the port of Coquimbo, and is the region's agricultural and cultural center. Cariaga, keeping his face toward the bay, doesn't concern himself to go up. Immediately, he dons his gear and jumps in to test the waters for himself.

"Alone in those two hours, I easily collected 50 kilograms with shell and all legal size (six centimeters). A man directed me to the bus over to Caleta Coquimbo's market. When I heard the price being offered, my dreams deflated. It was very disappointing: 25 percent of the San Antonio price (200 pesos per kilogram with shell)."

Cariaga returns to San Antonio and regales his wader friends with true tales of vast quantities, glossing over the poor prices. He also locates a buyer willing to travel back and forth with them, marketing the macha back in San Antonio.

"Our buyer drove eight of us to Coquimbo, and within the two low tides of a day, we'd filled his pickup and all drove home. But then our cannery wouldn't pay up for Coquimbo's smaller-sized macha! Our buyer was mad, and refused to commute with us ever again."

While the Coquimbo canneries were accustomed to processing their bay's macha averaging five to seven centimeters, the San Antonio cannery was dealing in their own macha, which averaged a couple of centimeters more. What appeared to be a wonderful commuting strategy flopped because Cariaga and his crew could not harvest a product that met the home processor's preferences.

Over macha chowder at his house in San Pedro, Cariaga continues his tale of how he ends up leader of Coquimbo Bay's waders. "We knew the macha were still plentiful in Coquimbo and that we could at least sell them locally.

Traveling to and fro failed, so we decided we'd have to leave our families behind for a while. Four other men wanted to return to Coquimbo with me, ready to camp on the beach."

He refills my bowl and continues, "San Pedro waders began studying us when they realized we kept bringing in loads more macha than they could. They'd gather on the beach where we'd gone in, puzzling each time we emerged how we could possibly collect such quantities. Soon they wanted our San Antonio secrets."

I wish to know the methods and distinctions, too. He says, "They started to call us *los mejicanos*—the Mexicans—because of our characteristic bandoleer, belts of lead weights slung across our chests and at the waist, and our extremely rapid foot work. We felt proud of the nickname and all the attention on the beach as we'd gear up, pulling on wetsuits, spitting in our masks to clean them, and tying long net baskets around our waists, the mouth held open by tubing hose.

"Armed this way, we mejicanos tap-danced on the bottom with our sensitive feet, able to submerge because of our heavy bandoleers whenever we felt possible macha. We'd look through our masks to check it out, gathering macha and leaving behind all the rubble of rocks and dead shells. Because of our method, we could wade much farther out into the waves than any of the San Pedro guys."

I ask how this had compared to the San Pedro waders' technique. "At that time," says Cariaga, "San Pedro waders stayed in their street clothes, maybe adding a couple old woolen sweaters, but without the aid of masks or weights. Staying shallower, if they felt it was a likely spot, they'd bend down, using their metal-wire net basket mouths to scoop up everything. It was a random, lousy method, requiring lots of energy for nada—nothing.

"Within only two days, word had reached one of the Coquimbo buyers, Mucho, who arrived at the scene. A beefy guy with a big gold cross, Mucho still offered us that low price by San Antonio standards, but we knew macha were plentiful. Besides, we could wade almost every day up in Coquimbo, weather conditions improving as you move toward the equator. And, weather that intimidated San Pedro waders was tolerable for us mejicanos. Out we'd wade, leaving the locals shivering onshore," says Cariaga.

Describing how Mucho attempted to monopolize them, Cariaga says, "Mucho rented our five-man team an apartment in return for us promising to sell him all our macha. We agreed. Beach camping isn't very comfortable and we needed a buyer anyway. Soon he grew greedier. Without asking, he delivered us his regular crew of San Pedro waders whom he'd been transporting to other bays near Coquimbo, and he expected us to teach them. Well, we did. But soon, because of the enlarged supply, Mucho tried to drop the price on all of us!"

Cariaga's face reddens at the memory. "We were just coming out of the water when Mucho dropped the bomb. The locals looked upset but swallowed it, knowing nothing about being organized. But we mejicanos were pissed off, having come all this way. Understanding the power of unity coming from a big caleta, we five looked at each other and I stepped into the ring. Right there on the beach I started giving a speech about having only two alternatives: either we got the usual price or we'd dump the macha back in the bay. Mucho didn't call our bluff. From this experience of facing the buyer together, I began coordinating all of us as the Coquimbo Bay waders.

"Feeling self-righteous, I returned to San Antonio to recruit more waders. Because of desperate straits there, I convinced almost everyone else to come join us in Coquimbo. Now we were sixty-five San Antonio waders strong."

I ask him how many of them now, a decade later, have settled in San Pedro. Cariaga says, "Just three, including me, actually shifted identity to San Pedro, developing deep ties. The rest moved on from San Pedro, first south to Region VII, and then continuing to migrate up and down the coast to find healthy beds. Whenever times are tough in Coquimbo, we go hunting for macha in other bays. I often cross paths with my old San Antonio friends, and we hang out together for as long as the wading is good."

That is the beauty offered by a migratory livelihood strategy: the ability to rely on shifting grounds and building social networks among fellow fishermen. This helps them cope with everything from overfishing to times El Niño events deal environmental blows.

Thanks to Cariaga's training and organizing skills that get Caleta San Pedro formalized in 1989, the waders start earning better income. They had enough product to draw in six buyers right up until the hiatus imposed by the terrible flood. As the macha recuperate, the buying system is resuming. The buyers circulate to each household, each family having shucked and bagged their macha.

Maximo, who had been a popular buyer, has recently resumed working in San Pedro. A barrel-chested man in striped sweaters now approaching fifty, Maximo's sociable, steady nature, taking time to drink coffee in his clients' kitchens, develops trust and wins him macha even when his price on a given day drifts a tad under another's offering. After helping shoulder several of Marlen's sacks of shelled macha into his pick-up, Maximo comes into the warmth of the kitchen where I'm already talking to Marlen.

She teases him, "OK, my Maximo, how much sugar in your coffee today? Are you paying full price for all our shucking I hope?" Maximo settles down at the table and starts nibbling on a sweet biscuit and smiles. With macha much in demand, he and his sellers have become allies in the flux of access to macha and in costs and profits in what has become a wide-flung operation.

Maximo, like some other buyers, maintains connections with truckers and processors as far away as Quellon, on the southern tip of Chiloe, Region X.

Marlen's macha meat will be transported that far over the next few days. However, she and her kids stay put, holding down the San Pedro home front even when her wader husband has needed to migrate to support them.

As Caleta San Pedro wading grew, it was no surprise that divers and waders were beginning to compete for macha as had happened in San Antonio. The same year the waders formed their caleta under Cariaga in 1989, the local diver fleet was already forty boats strong. At the wave-break zone, the divers started edging into shallow water while the newly San Antonio-trained waders were now able to venture deeper, doubly leaded and suited up.

All during the 1990s, the two groups were overlapping and impinging on each other's grounds, diminishing the beds from both sides. By the time the flood hit Coquimbo Bay, choking out all beds, the dive fleet had redoubled to eighty boats. In the face of complete natural ruination, the divers and waders were left arguing over who was more to blame for the previous decade of depletion.

During the winter months from 1992 through 1996, about thirty San Pedro waders led by Cariaga migrated northward in the hopes of better macha income. Camping in other Chilean bays, they reached to Huasco and Mejillones. Remembering to what lengths they had to go to make ends meet, Cariaga launches into the story of how San Pedro waders successfully migrated and set a precedent for waders remaining outside artisanal fisheries law.

"In 1995, we made it all the way up to Arica, the last coastal city before the Peruvian border. And who did we find already settled in for several months and busy training a mass of 500 Arica locals? Our dear San Antonio wader friends! They embraced us, but Arica's mob wouldn't let us stay, not another thirty outsiders that they didn't need to train them.

"So I wrote to the Chilean Navy commander posted on the Peru border, requesting permission to harvest a beach abutting the frontier. The commander agreed as long as (1) we were mindful of the land mines, and (2) nobody had a police record," smiled Cariaga.

"I attributed this success to my polemics to the commander: Where does Chilean territory go to? With our success, the Arica waders got jealous and asked our San Antonio friends how on earth San Pedro had won this unheard-of permission from the Navy. But never mind," Cariaga said, "For the next two years about one hundred of our San Pedro waders came up and joined in. We were renting houses only half a kilometer from the border in a tiny settlement of llama herders near the Arica airport. And did we ever harvest macha!"

Regulators really viewed waders as outside the purview of artisanal fisheries management. Despite the 1991 Fisheries Law attempting to anchor fishermen and end migration, the government continued to let waders migrate under a de facto Chilean policy: nobody is owner of the beach. Waders announce to anyone, "No papers are held by any other locals to stop us, so we enter freely."

Cariaga says, "These are the same reasons we waders proudly claim to be the last true artisanal fishermen, heirs to the pre-Columbians. We think of ourselves as free spirits above the government registration system."

This became the sticking point. The National Fisheries Service—the agency in charge of enforcing fisheries regulations—hardened their line against San Pedro's waders, expelling them from the border and sending them packing back home.

At about the same time that this government hammer dropped, the particularly powerful 1997–1998 El Niño event began brewing in the southern oceans. Within the year, waders in Coquimbo Bay left in the wake of the flood confront not only the burial of their macha, but also sharpened antagonism from divers, regulators, and marine biologists from Universidad Católica del Norte, who view waders as problematic. Biologists perceive danger in continuing to condone waders in Coquimbo Bay. This is because of Coquimbo's type of macha bed.

Macha live in two distinct types of beds, depending on the degree of beach slope. However, all juvenile macha recruit—hatch, if you will—in the shallower waters, migrating deeper as they mature.

Tongoy Bay has the first type of bed, one that's better at tolerating waders. It lies exclusively inside the intertidal or surge zone, reaching a maximum depth of 1.5 meters, where only waders can harvest. Here, mature macha never migrate out past the surf zone, so waders work on the entire population, tending to target adults in the deeper areas near the wave break.

Coquimbo Bay, like San Antonio, has the second type of bed, gradually extending from the intertidal zone out to a depth of five to six meters, far beyond the surf, so that both waders and divers share the same stock. Juvenile macha stay in the waders' area, between the coast and the surf zone. Only as macha mature will the majority move out to greater depths where divers search the bottom.

This explains why university biologists see Coquimbo waders as more destructive than divers. Waders here end up with a 70 percent juvenile catch, plus the few stray adults who remain in the shallows. Given Coquimbo's type of bed, one can wonder if waders can choose a conservation ethic, despite any good intentions and leadership.

As the waders' leader, Cariaga has had to learn the important interrelations between divers and waders in terms of fishing effort and impact on the resource, as well as organized political power. Cariaga parries the biologists' negative view that waders injure the juvenile stock by saying, "It isn't hard to realize how divers harvest significantly more than waders. We waders have natural restrictions, being dependent upon two daily low tides when wading is possible. Divers using boats have the option of working six hours a day, plus they can tolerate worse weather being deep on the bottom, far below the surf.

"If there were no divers, waders couldn't possibly wipe out the bank because it extends far out past our reach, beyond the break," says Cariaga. This seems a fair assessment for the sloping banks of Coquimbo and San Antonio where, by comparison, the divers can harvest more intensively than waders.

In his closing defense, Cariaga wants to impress upon me how almost all divers as well as waders will ignore restrictions, such as taking whatever size macha the market will buy to earn a living. "The market decides the sizes we can bring them, not laws," he says, which is another common pattern. Reflectively, he adds, "It doesn't matter whether we're waders or divers. None of us think about tomorrow's bread. We look at the present conditions."

I love this popular expression: *No piensan en el pan de mañana; miran el presente.* (Don't think of tomorrow's bread; look at the present.) Being caught up in the moment reflects fishermen's sense of immediacy in deciding how to support their families—livelihood decision making. This is globally common among fishermen, who decide what, where, when, how, and with whom, based around immediate survival. Fishermen's focus is on locating that "bread" for the household, in light of their moment-to-moment understanding of oceanic changes, the unpredictability of the sea's bounty and sudden risks to their own safety, let alone survival.

Angel Cortez and René Maluenda Spearhead Divers' Management Areas

The two caleta presidents, Angel Cortes of Caleta Coquimbo and René Maluenda of Peñuelas, have served together in their Region IV artisanal federation, and maintained a close friendship over the years, helping them jointly develop macha Management Areas—*areas de manejo.* A common denominator among fishermen attempting to manage their fisheries is the presence of a charismatic leader who pushes to implement directives. Like Nelson in Puerto Aldea, Angel Cortez and René Maluenda are such men. Through them, the dive caletas begin shouldering self-regulation, despite two beliefs commonly held by fishermen that preclude strict self-management: (1) management regulations can be broken when they hinder earning a living; and (2) government interests primarily favor industrial fisheries and coastal tourism, the real culprits of pollution and diminished resources.

I go to visit the long-term president of Caleta Coquimbo. In power since 1983, Angel Cortez is a local, self-educated artisanal fisherman. I wend my way to the Caleta Coquimbo office door, enjoying the delicious walk through their seafood market, tasting the colorful cups of raw shellfish in lemon juice. Passing through the gate into the caleta's boatyard, I'm fascinated by women deftly hooking individual baitfish onto each longline hook. It's a tricky procedure if the line is to keep tangle-free and able to uncoil rapidly at sea. Pelicans are rafting around the boats along the bulwark, jostling and flapping after scraps from men offloading their day's catch. We're surrounded by the constant sounds of scraping and painting of wooden boat bottoms. A guard in a heavy hand-knit sweater opens the door of a simple towerlike building, and leads me up dark stairs and knocks.

Cortez, stocky and powerful in nature, comfortable in himself and with his command, welcomes me in and orders the man to go bring us coffee. He sits back in his desk chair at the far end of the room, and I settle in front. First I explain how I'm working up and down the Chilean coast with fishermen, focusing on macha fisheries here. Then I begin asking his work history. I want to discover caleta leaders' strategies for understanding and dealing with their vision of change and management.

Cortez says, "Chile's deep recession of 1981 to 1983 was awful. Fishing was all most of us could find as work in Coquimbo, including me. I'd been forced to return home after ten years mining in the altiplano when that had been the more lucrative work." Mining to fishing is a common seesaw pattern among Chilean workers, one we also see in action in urban Iquique (chapter 6).

"We knew how labor could organize at the mines, and saw how Pinochet was punishing unionizers," says Cortez. He had become politicized, like many who throw themselves into organizing caletas. "Once back home, I was determined to get fishermen some kind of control by organizing so we wouldn't lose even this life. Realize that Caleta Coquimbo is much more than macha divers. We are divided into eight separate fishing arts, up to 400 members overall.

"Hearing of Peñuelas doing great with macha diving, divers moved here, so we launched our own macha fleet. University biologists began crying that our good level of collection—500 kilograms per boat per day on average—was hurting the macha, yet we didn't agree. But by 1986, we realized that macha prices had fallen because quantities were too high. So I held meetings with our divers and got them to understand and agree to bring in less to boost prices."

"Two years later, 1988, biology advisers of the highest minister for fisheries came to the bay to start research on macha management. They used our divers and boats, in the end recommending harvest cuts. This spawned our interest in Management Areas.

"Biologists told us how participating in Management Areas would give caletas rights, so I saw how it was pivotal to position our macha divers with a Management Area. I decided to team up with the university biologists and managers and Peñuelas to collectively regulate a proposed Management

Area. This meant strictly imposing self-regulation, which I explained to our divers in terms of claiming rights to the area."

We sip the energizing coffee sweetness. I knew from others that Cortez is the respected chief of the biggest caleta, and it was easy to sense how he could convince assorted fishermen, and assert leadership to balance political rivalries.

He continues, "Politically, I must apply my authority to coordinate all three of the bay's caletas in making Management Areas, even if those San Pedro waders don't come to meetings. I am compelled to oblige them to obey or else we'll lose the Management Area bid."

I understand his reputation for not delegating, his strength of character radiating, he's proud of being personally in charge of the creation of all aspects of the macha Management Area. We hear footsteps tripping up the stairs to join us. It's René Maluenda coming through the door, the president of Caleta Peñuelas, here for one of his usual conferences with Cortez. In contrast, he's younger and still fit to dive the cold bay currents. And Maluenda controls a small, tightly knit caleta community.

After Cortez recaps our Management Area discussion, Maluenda speaks with light in his eyes and a nod to Cortez, saying "I'm more a negotiator than a strong-arm type. I get my points across by speaking softly. I know I'm unusual in this, whereas 'my wife' runs our potential Management Area with a big stick."

His sexual joke reflects how closely these two can work, evoking laughter all around. He grins at me. It's true that his sweeter method is a rarity among caleta leaders. But Maluenda and Cortez complement each other in pushing their divers toward self-regulation and participating in their own Management Area management. They generate faith and trust among their fishermen, enabling them to take political-economic actions.

Yet, like many leaders, they must endure a level of mistrust and a lack of cooperation from some fishermen they represent. It's standard for fishermen on the fringes, whether inside or outside the organizations, to grouse or fight against leaders' proposals, such as for self-regulation. Sometimes these fishermen accuse leaders of personal benefit—as in Iquique (chapter 6). Leaders often agree with managers in hopes of gaining some input in impending regulations. This may meet with categorical opposition from those fishermen wanting to continue ignoring a regulatory system. A common global pattern in fishermen's organization, it's one that I've witnessed from the Florida Keys (Meltzoff 1997) to Galicia, Spain (Meltzoff 1995), Chile, Peru, and Galapagos being no exception. As liaisons between government authorities and university biologists, and the caletas, Cortez and Maluenda must negotiate resistance from both sides to reach a mutually acceptable degree of participation in management.

In Chile's Management Areas system, a caleta could legally begin guarding its Management Area once it makes a formal application. This requires having a map drawn delineating the Management Area location and signing a decree before the costly basic biological study can begin. The dive caletas selected and set aside the most stressed areas with the fewest macha as proposed Management Areas and began guarding them. By 1993, the dive caletas had set in place quotas and fines, stiffer than legal ones. Their daily dive quotas were not based on biological data, but established on livelihood economics: how much could satisfy their daily needs and sense of well-being. They toughened the quotas over time, leaders urging ever tighter control as caleta initiative. They achieved this by restricted harvest days and amount per boat. However, there was nothing about the number of dive boats that could work—called "limited entry" by managers.

René Maluenda says, "At first, we limited harvesting to five days a week, and then later to just Mondays, Wednesdays, and Fridays. And the exact areas we'd mapped, being the poorest ones became off-limits."

Government dive quotas for legal-size macha also shrank, but harvesting was still heavy. To the managers', biologists', and divers' distress, waders continued their heavy unregulated fishery for the juveniles. Divers, more than ever feeling the weight of strictures, were pointing at waders as the worst, most persistent, unchecked perennial problem. Caletas fined divers for juveniles brought to the wharf. Divers knew they were supposed to spend time sorting out juveniles on board, keeping them alive until they reached the Management Area to release them for repopulation.

One diver summed it up to me after he was caught red-handed with juveniles at the caleta wharf. He swore, "But macha grow up on the waders' side of the waves. That's what our biologists explained. So waders always take them. What good does it do for us to be so careful?"

Maluenda explains, "At heated meetings, Cortez and I had the juvenile conflict thrown at us. We had to remind divers that two wrongs don't make a right, and that they had to show good faith in self-management to win Management Area status. But, because beds were dwindling from waders' illegal harvesting, our university advisers kept pushing us toward greater self-restraint. We could listen and act."

For several years before the flood, with the steadfast support of university biologists, divers begin organizing Management Areas without the desire or political will to include waders. The divers are "othering" the waders—seeing them as separate and less worthy than themselves. Out of what the divers considered their own good graces, macha Management Areas informally include the waders. Dive leaders halfheartedly try to include the waders of San Pedro in planning, inviting leaders to meetings that they rarely attend. Relations are weak at all levels. From the waders' perspective, it is not a secure position for their future. From the perspective of divers, waders are annoying and prove difficult to incorporate into macha management since they're not one solid community. Jobless people wade in when hungry.

Management Areas are not physically isolated from their urban milieu; all three caletas have had to deal directly with men, women, and children from poor urban neighborhoods, who often wade for subsistence. There's no reason spontaneous waders would give any thought to government's legal size limits and registry regulations—the only two regulations that a de facto Management Area could invoke.

Lumping San Pedro waders together with the poor's happenstance waders, managers view them as dependent upon the macha for economic survival. Managers call them *casos sociales*—welfare cases—and look the other way. Macha is serving as a kind of welfare subsidy.

The university biologists have little sympathy for waders since they deplete juveniles. Charitably overlooking illegal fisheries or bending management to accommodate hard-luck cases, such as the waders, does not resonate with their scientific perspective. Nobody could expect waders without biologist allies to carry out costly studies, let alone control a flux of outsiders in the wading zone, or to be able to organize themselves to meet requirements for Management Areas. Hence, waders, holding the least power in Coquimbo Bay, fall into the category of political outcasts from the Management Areas system.

Without waders' involvement, the dive leaders proceed. Cortez and Maluenda each lead their caleta in devising a strategic plan for divers composed of three commissions: administrative, social, and fiscal. Then their two administration commissions meet to establish and coordinate accords between Caleta Coquimbo and Caleta Peñuelas on quotas, reserves, and fines, quotas being the principal management method employed. They decide which areas of the Management Area to declare as reserves and which ones to repopulate. Another of their management methods inside the Management Area is to allow divers to "weed" the plentiful *taca*—a small clam that's the macha's ecological niche rival. Taca harvesting is permitted on macha "off" days.

The social commission's role is to cover sick pay once they verify that a *machero* is ill, and costly funeral expenses if he dies. The fiscal commission enforces accords at the caleta level, posting observers on the water and at the caleta dock to check for compliance with quotas and ensure against the collection of juvenile macha.

Cortez is hands-on and not a delegator; he heads each of his caleta's commissions. Cortez personally shows up at the docks to monitor offloading. If a boat brings in more than its quota, then the caleta confiscates the surplus to sell, the revenue going into a general fund to be divided between everyone at years' end. This regulatory method also relies somewhat on peer pressure. "We taunt someone for being stupid and overworking if they bring in over the quota," says a Coquimbo diver.

Caleta Peñuelas, however, relies on a far more casual system of enforcement and peer pressure, as they pull each other's boats up on the beach to offload in front of the caleta in plain sight of all. A Peñuelas diver says, "We'll take away fishing rights from anyone trying to hide macha on the beach."

Excluding the waders, the dive caletas finally propose two macha Management Areas in Coquimbo Bay, making an accord to share them, although technically, Peñuelas is the owner of both. They agree to open harvesting three days a week. The macha banks of one of the Management Areas rest in the shallower, sandy-bottomed areas. Without input from Caleta San Pedro, it tantalizingly runs for six kilometers directly in front of San Pedro, from the La Serena lighthouse north to the bay's point, and from the shoreline out. Moreover, the Management Area's inner shoreline fringe has been the historic stomping grounds of the San Pedro waders. The other Management Area sits in front of Caleta Peñuelas's beach. Its sandy-bottom bed stretches more than 1,000 hectares, running long and shallow for about eight kilometers across the bay in up to ten meters of water.

Designed as a diversification strategy for the caletas, there is even a third Management Area for a separate non-macha dive fishery, yielding scallops and *caracol rubio*. Although the largest in area, it is considered poor quality. It lies in deeper water near Caleta Coquimbo, running down eight to twenty meters with a muddy bottom that has always been too turbid for divers to harvest successfully, or for biologists to evaluate. The hope is to communally harvest it several times a year at best.

Initial results of self-management impress both divers and university biologists: much bigger macha that sell for over five times the price of those usually harvested at minimum legal size. The biologists continue to push for ever-more stringent quotas. Cortez comments, "The national market price seemed just too low for us to pass fresh quota cuts along to the divers. But, our biologist friends convinced us to go ahead and trim back for a year-long project monitoring the bay."

In this monitoring year where divers exercised great restraint, they ultimately cut daily harvests by about 80 percent, restricting diving time to three specific days a week. Divers have realized that if they need more time to harvest the same amount of macha, the scarcity is probably from overfishing. They therefore agreed to curtail fishing further.

Ironically, the meaning and impact of their conservation actions came into focus on the eve of the flood, which will make it wasted effort. "Years of self-management for naught?" echoes in many minds as the macha silt to death. "How will this play out for future willingness to restrain themselves?" wonder the university biologists, like Wolfgang.

As the 1997–1998 El Niño event unfolds along the coast of Chile and Peru, it brings unpredicted environmental devastation as well as an abundance of other marine life. The Coquimbo Bay case is devastating as sediments from the river suffocate all the algae and almost all the filter feeders, including the macha. During the rains leading up to this bizarre flood of August 1998, neither divers nor waders could possibly realize that a narrow window of one month remains to salvage the macha. For many divers, almost a decade of self-control is about to transform into a quixotic exercise. René and his fellows watch helplessly from the Peñuelas beach in the pelting rain. The general runoff builds into such murky bay conditions that collecting becomes impossible. They have no contingency plans for an emergency harvest.

With the rapidity of events, René says, "Macha succumbed in under a week from a sudden blinding turbidity and surge that made any attempt by us to salvage macha too dangerous. We stood alongside our boats onshore, feeling crushed, especially after all our sacrifice, and wondering what fishery can sustain us now."

Nobody had predicted this specific El Niño scenario. Neither the press nor the fishermen nor the biologists foresaw the Rio Elqui flood with its deadly outcome. Caught off balance, the *macheros* had never dreamed all their macha could be wiped out. In their concept of natural resource use, it is wasteful to let macha perish unharvested. They were ruing their inaction, not having collected the macha before it grew too risky.

There was no local knowledge or memory of this kind of circumstance. The El Niño event of 1982–1983 had benefited macha. Shellfish beds recovered brilliantly with the warmer waters triggering great spawning, combined with driving off most predators, which are unable to tolerate the heat. There had been no flooding, just storms. The truth was that in the 1997–1998 event, the situation had grown far more unstable with agricultural development along the Rio Elqui.

Post-Flood Waders' Predicament

The El Niño flood also wipes out wading. Simultaneously, the government tries to clamp down on waders, to bring them under regulations and attempt to incorporate them into the divers' fledgling Management Area. Before El Niño, 40 waders from San Pedro had formally registered. After El Niño, all of San Pedro's 153 waders put their names down on the government's regional list for macha, signing in as *recolector de orilla*—designated coastal collectors. Now of all times, waders feel the need to migrate to find other grounds until the bay recuperates. They're motivated by an accord with neighboring Tongoy waders, who let waders come there as an alternative with the caveat that they register. They desperately need this livelihood strategy. The rules being imposed on waders actually forbid migration outside the home region. Tongoy is nearby, but many other sites they have relied upon in the past lie beyond legal range. In general, however, rules are ignored when they inhibit survival.

In essence, waders think they are receiving the brunt of poorly executed fisheries management politics. Cariaga sums up the waders' perspective on why all of a sudden they are the focus of regulation, "For years we've been viewed by biologists and managers alike as an environmental danger, yet we were tolerated. With the wipeout, toleration ends. Given everyone's frustration at ruination, we become the scapegoat; and they fear our numbers on the loose."

Coquimbo Fish Market without Macha

October 1998, I'm sampling the fare at Caleta Coquimbo's bustling Sunday morning seafood market. There's nary a local macha to be found. Women and men seated behind their low tables hawk their heaped portions of fleshy shelled shellfish in whites and yellows rimmed in reds and purples. Macha are mixed in, but when I ask, all the macha and the other shellfish have been brought in from surrounding bays that didn't flood. It's a family crowd from the urban centers bordering the bay (Coquimbo and La Serena) hunting for their Sunday brunch. Besides selecting glistening fishes with fresh red gills and still bright eyes to cook up at home, grandmothers to little kids are enjoying paper cones filled with a mélange of shellfish. Others are sitting down at mom-and-pop concessions at the fringe of the vendors. Plates are heaped with seafood. Establishments with striped awnings line the bayside as musicians stroll from table to table, strumming popular requests. Today, I choose an umbrellaed table right by the edge overlooking boats swaying at anchor. A female sea lion is swimming up and down the bulwark, her face and whiskers above water as we exchange looks.

After lunch, I wander inside the working caleta and spot Jaime, one of the former macha divers trying his hand as a baiter alongside those who usually perform this piece work. He's not-so-dexterously untangling masses of longline to help reorganize the barbed hooks on boards for rebaiting and the next fishing trip. "It's temporary," Jaime says, and smiles as I shake my head in awe at the puzzle.

Luis, one of the waders now out of work after the flood, is a slender young man who, remarkably, could resist the cold without his own blubber layer. Luis confides in me how before the flood he and several others from San Pedro had been swimming out from shore into the deeper waters and attempting to free dive—holding their breath without any hoses or tanks.

"Really?" I ask, surprised.

Luis tells me how they felt the quantity of macha had already been spiraling down, stressing, "We weren't going to let the divers be the only ones crisscrossing zones. Switching to snorkeling in bottom currents and surf on the surface was hard. Anyway, it could never have sustained San Pedro waders. There really weren't enough macha if all three caletas were harvesting beyond the break."

I ask, "What do you plan to do now, for as long as it takes the macha to return and mature?"

He nods slowly to himself, saying, "About forty of us from San Pedro are heading down to the southern tip of Chiloe next week, but the women will stay here at home." Luis would leave wire thin already. But, upon return many waders will tell me how they'd left fat and returned skinny, literally and figuratively, from the cold and hunger of camping and wading in Patagonia. Compare their short-term experience to those of the migrants colonizing the first Patagonian merluza camps (chapter 3).

Unlike Luis's group, who willingly withstand a stint in Patagonian waters, another dozen waders feel lucky to find local employment widening the highway south of La Serena. Roadwork is one of those unskilled, temporary options for fishermen. But most must go find cash elsewhere.

Family Martin Wading outside Coquimbo

Family Martin lives behind San Pedro's giant dunes. A seasonal gypsy group has set up an encampment near the caleta. Coming in, I admire a huge blue-and-white-striped circus-style tent in the morning sun reflecting off the high dunes blocking them from the ocean winds.

When I mention the gypsies to Maria, the mid-life mother in her long, tented housedress belted in at the waist, I find her feeling far from indulgent toward their transient neighbors. She might have been expressing her own family frustration at being banned from wandering in search of macha, besides expressing social prejudice toward gypsies, who perpetually shift location, insular among themselves, living outside social norms. "Those gypsies must go away," Maria says. "We can't spare anything and they're a burden if they want to find food here."

I feel an irony surrounding the migratory lifestyle. It's hard to accept the necessity of "others'" (non-waders, personified by gypsies) migration as a lifestyle. Maria continues, "Look at us! We're going to have to go soon. We'll try farther up the coast where we've camped before, building a *ruco*—rustic patchwork house—on the beach for a few months with several friends from here. Law has no teeth to keep us hungry."

As Maria vents, she is stirring up a big dented pot of potatoes and beef bone, dishing it up in blue cracked bowls for her husband and me. Her married daughter, Raquel, is ironing by the kitchen table. Raquel says, "I'm staying here, Ma. My man has already gone without me. I hate rough living on the beach, cooking on fires and forever defending the windward side of the *ruco* with scraps and boards, or else it's just too cold to sleep." For my sake, Raquel adds, "Besides, his group is there without women. I want him to come home with enough stew money to tide us over for several more months together here."

The majority from San Pedro are intent on sticking relatively close to the comforts of home. Therefore, about two months after the flood, they opted for driving the hour south to wade in Tongoy Bay. Being in the same Region IV, this isn't illegal migration in the eyes of the newly vigilant post-flood managers.

Pepe is cool, about thirty, hair cropped, dressing 1990s "Euro trash," his upper body in tight T-shirts, sleeves rolled up with a pack of cigarettes tucked into one side, and sporting shades. Pepe tells me, "Tongoy's only a short distance and some of us had friends from over the years."

When I ask how he thought the Tongoy Bay waders would react, Pepe laughs, "We realized that about fifty San Pedro waders, plus another twenty poor strays, all marching into Tongoy Bay, couldn't make us popular. But being desperate, we felt justified."

Actually, they descended on Tongoy's Management Area. Back in 1991 during the advent of the Management Area system (Meltzoff et al. 2002), a federation of all of the region's caletas—including Caletas Tongoy, Puerto Aldea, Peñuelas, and Coquimbo with San Pedro—created an Accord of Respect among themselves to acknowledge each other's Management Area rights. It didn't matter that their government Management Area status was in limbo. But Management Areas and their status were not primary in the minds of San Pedro waders. Waders have had that vital and separate ideology of allowing a social security safety net.

"How was your gang received the very first day by Tongoy waders?" I ask.

Describing the fight, Pepe says, "As we were starting to come out of the water with our macha, some Tongoy guys spotted us and starting shouting insults at us to scram. They were pissed off as they came upon us. I punched a guy back, as did my buddies. It got rough, but we still left with our macha."

I wonder if there was any talk, like of their plight. "No," says Pepe, "Those guys all knew from the news we'd gotten wiped. Besides, there's pride and ethics. We would never go into their turf just for the thrill or for extra macha."

Ill will brewed, partially because the Tongoy waders' leaders were powerless to reign in Tongoy's random independents, and they knew that San Pedro was equally powerless over those coming with them from Coquimbo area. Both classically put up with lots of strays who were not caleta members.

Ultimately a solution arose through the compassionate intercession of some Tongoy wading women. The women reminded everyone of the dire straits in Coquimbo Bay, and how they, too, may need reciprocity in case of natural disaster. This elicited a level of empathic treatment.

Intoning the logic of the classic migratory system, Gleti, one of the outspoken Tongoy women who stands up for helping fellow San Pedro waders, says, "We could be in Coquimbo's place next time. The sea doesn't show favorites, and we don't control her whims."

Thus the clashing waders resolve their conflict with an accord between Caletas Tongoy and San Pedro. In November 1998, Tongoy grants San Pedro the exclusive rights to harvest a two-kilometer section within their Management Area's macha bank until the Coquimbo Bay macha recuperate. In return, San Pedro leaders accept responsibility for their waders in (1) honoring the legal size limit for macha and (2) legally registering, a strong incentive so that all finally comply. Now it's up to San Pedro leaders to develop and enforce their own mini management plan for their section of Tongoy's Management Area. "Maximum sustainable gain" in biologist-speak becomes a goal.

In less than half a year, by mid-1999, Tongoy waders grow fed up with San Pedro, thinking they were cheating and taking all sizes of macha, and sacking the Management Area of as much as they could. As a strategy to end the harvesting accord and kick out their unwelcome guests, they ask the National Fisheries Service to shut down the whole Management Area. This provokes an unexpected result. Once Tongoy's total Management Area is closed to any caleta waders, scads of strays in street clothes wade into the void. These temporary waders start taking advantage of minimal supervision, stimulating the next free-for-all.

"Why protect the area if, legally, nobody can harvest it," asks the local street vendor, Jimi, who usually sells me envelopes and postcards. Acting on his line of logic, Jimi, in green nylon shell and clinging wet sweat pants, hair streaming and fresh out of the water, is piling up his macha on the hard sand.

The National Fisheries Service again steps in, this time asking the Coquimbo University biologists to reorganize the Tongoy situation. The biologists tell me they feel as though they're perpetually working on Management Area studies but without succeeding in legitimizing a single Management Area, including this Tongoy one.

I ask Pablo, one of the biologists, how they've been dealing with the pandemonium over macha. He explains, "With the new authority given to us, we've reestablished limited entry, for only Tongoy waders, capping their number at twenty-five, with correspondingly higher individual quotas for macha. We want to make waders' monthly income worthwhile, so they'll become territorial and guard their Management Area. Of course, our efforts will be stillborn if the Management Area sits pending."

Post-Flood Divers' Strategies

Among Coquimbo Bay's divers, migration initially was their key to survival after the flood. While waiting for the bay to recover, many divers delve deep into southern Chile's distant macha and loco grounds. Despite migration between regions having been outlawed (1991 Fisheries Law) Chile's marine enforcement arm was showing sympathy toward the divers, turning a blind eye. Upon the divers' return, I asked some of them about the famed proclivity of Region IV fishermen to migrate. The poet among them, Martin from Peñuelas, explained, "It's a natural practice for us, living right in the center of our beautiful strip of a country."

Throughout the 1980s and early 1990s, Martin reminds me how Peñuelas divers periodically migrated for months at a time to harvest high-priced species, such as loco in the northern regions, and loco and merluza in southern regions as far south as Aisen and Canal Moraleda in Region XI (chapter 3).

"Our movement is nothing new," says Martin, forgetting the others around our table, eyes shining in the low yet intense storm light of a coming thunder. Conversations often turn quite intimate, people leaning in closer to share passion for this sea life, the challenge well met, the excitement of the unknown. Policy makers and processors may tout policies and projects to stabilize fishermen's income through settlements, fixed markets, or aquaculture. But, the fishermen don't believe in it or are not concentrating on something that might not pan out.

"The first good break after the flood came five hours from Quellon, off Achao beach. Exploring from two open boats, a few of our *macheros* located a bounteous virgin bank. Back in Quellon port, they sent us a *convenio*—informal words among friends—telling Coquimbo Bay divers to come quickly since this has the potential of a mini-boom."

Martin describes the impressive response, saying, "Caleta Coquimbo divers trucked three boats down to Quellon, but we Peñuelas divers formed a giant 150-man crew, bringing thirty of thirty-six dive boats, just about evacuating our home beach. Our buyer, Tapia, came with us. We set to sea in Quellon and dive Achao over the next couple of months. Waves were rough, and with winter nearing, sea conditions only let us dive one out of three days. Summer weather might have brought twice the water time, but the harvest was fat."

One young Peñuelas diver at the table, Donato, sums up the stamina for Patagonian diving, saying, "I kept at it for three months, despite the bitter cold, but the real macho diehards—twenty strong—were still there a year later, sending home news that they'd rather be wet than come home to the dry life of a macha-less bay."

Migrant Camp of Divers during Tough Times

Only men had migrated with Tapia, no women. One of them, Rudi, is cozily back in Peñuelas, sitting in a local seafood restaurant describing the unforgettable sleet of Achao camp life. Rudi's blue eyes lower, and he tugs his sweater tighter around his neck. He says, "We set up cheap nylon tents that leaked in the perpetual drive of rain and ice. At times the supply launch couldn't make it through the surf so we had to go hungry. Men cried. No dry sand here, only mud."

The door swings open, and another diver passes under the decorative relic of a wooden row boat suspended from the ceiling. He comes by our table with warm handshakes, heading to the kitchen to chop. We order more seafood empanadas, and Rudi continues, "Our secret spot leaked to Quellon men, who saw the catch being landed. But they didn't know how to dive or have any gear. Maybe a thousand guys move in on us to wade in close. They'd edge out into the water, shivering without wetsuits, wincing at each splash, but the cash was too tempting."

"The mass of waders comes as a blessing for Tapia and us," says Rudi. Nailing it, he says, "Buyers legally can only purchase from regional waders, but not from us, unregistered from Coquimbo. But, once onshore and back in the Quellon market, who could tell who harvested what? So the whole catch becomes legal, a big advantage. The mini-boom was too short for regulation to descend."

After empanadas, I say good-bye to Rudi, and walk along the yellow sand shore, cutting up a hill road to Lapita's house. I knock on the door; his wife answers, telling me to return later that afternoon. When I return toward evening, Lapita explains how he, like most buyers, operates. He says, "I weigh potential profit against the risk of receiving fines for working with my fishermen, who're unregistered in all other regions. Our Achao mission turned into a boomlet, with rich hauls by my Peñuelas group, then the rush of hundreds of inexperienced waders desperate for cash, camping as best they could and collecting whatever they could, with lots of buyers milling around in the riot of emptying the bank."

When Peñuelas macheros migrated en masse to Achao's banks, Angel Cortez of Caleta Coquimbo stepped in to cover the politics of the home front, like the Management Area. Within the year, most Peñuelas divers return home with cash in their pockets from the grim muddy reality. They turn to odd jobs, urban opportunities in construction or restaurants.

In contrast, only a few macheros from Caleta Coquimbo had ventured to Achao. Yet one-third shift over to join Caleta Coquimbo's sector of deep divers to keep employed, opposite their Peñuelas fellows. Only three of about two hundred Peñuelas macheros opted for this transition into gloom. I feel like the macheros when it comes to diving conditions. Deep divers usually sink well below the legal limit of twenty meters, entering dark waters to spear fish and hunt the precious shellfish living around rocks, such as loco, chocha, and piure. Although Peñuelas macheros are registered to target these other species, as is common among fishermen to insure flexibility, they feel that their gear, skills, and comfort zones, are incompatible with such endeavors.

Macheros repeatedly tell me how they prefer sunlight streaking through the water in the less-pressured shallows. They don't wish to tangle with the dangers and difficulties adjusting to atmospheric pressure and blackness. Diving in deeper conditions would also require a change in familiar gear, especially the method of weighting oneself. Whereas macheros are accustomed to donning lots of weight around their ankles and on their backs to hug the bottom in the surging shallows, deep divers keep their lead weights tied at the waist, like scuba divers normally do.

However, some Peñuelas macheros do begin a completely different fishing art: longlining at great depth at sea with baited lines. Within months of the flood, Peñuelas's captains and crew, of the fourteen-boat gill-netter fleet take the lead in training their friends in the six remaining caleta boats that hadn't flocked to Quellon to fish alongside Caleta Coquimbo's deep longlining fleet. This post-flood stint fishing is illegal, but this is not one of their considerations. In switching to longlining, these macheros consider the work much more of a physical sacrifice than diving. Five days a week they must set out at 2 AM into the pitch-black cold, returning to their beach at noon. Whereas for macha—given the restrictions of their caleta and the tides—they could only dive two or three hours, just three days a week. For longlining, moreover, the women are needed for the tricky job of preparing the bait. So, women learn to string the lines out on Peñuelas beach, baiting each one of myriad hooks tied along the meters and meters of heavy nylon line. Once offloaded, these lines have to be untangled and rebaited for that very night's trip. They can sell the catch of good-sized (1–3 kilogram) hake in the capital's national market, making it all worthwhile. This hake—a different species of hake from the merluza austral (chapter 3)—at first proves a successful avenue for Peñuelas divers awaiting the return of macha.

Peñuelas divers throw themselves into working hard at learning and investing in longline gear, despite the lack of official permits. Their gamble proves correct. Because of the temporary, dire circumstances, the captains and crews of Caleta Coquimbo's fleet of forty-six registered hake boats grudgingly accept them, just as do the National Fisheries Service managers who look the other way despite the closed registry.

Thus, the Coquimbo fleet shoulders the addition of illegal newcomers, reassuring themselves it's a short-term gig for the divers—a fellow fishermen's courtesy and social safety net—an interim survival measure while awaiting the return of macha. These thoughts soften the fleet members' urge to blow the whistle, calling down enforcement from the National Fisheries Service.

But soon, the longliners start hurling recriminations at Peñuelas because (1) the government quota finishes early and (2) the increased supply of hake on the market was letting buyers slash the price in half by late 1999. Everyone definitely charges ahead fishing after the quota is filled, goading each other on. But the National Fisheries Service managers continue to turn a blind eye at this latest flagrant act.

Managers feel justified in viewing the hake fishery as a sacrificial outlet to foster fishermen's well-being. Managers don't have the heart to stop or fine any hake fishermen that harsh year. Their compassion toward transgressors in the maw of environmental collapse and poverty comes from the concept of using fisheries as an economic and social safety net. Managers commonly have made conscious exceptions, and now Coquimbo divers and longliners, as well as waders and their strays, all become entangled inside the web being granted clemency.

Macha on the Brink of Return

Everyone—from managers and biologists to divers and waders—eventually expects Coquimbo Bay to recover because of the swift and rich ocean currents naturally feeding and seeding the waters of the Chilean coast. But some managers worry that it might take years for the macha stock to fully recuperate from total loss of stock and habitat.

Divers and waders stay more sanguine. They think that although Rio Elqui deposited dense deadly sediments on the banks, macha seeds are always sweeping in from the sea, and water action constantly shifts bay bottom surfaces. "Before long, our macha will reappear and we can dive here again," says Nado as his diver buddies nod.

Sure enough, compact deadly flood sediments begin dispersing in El Niño's inordinately rough seas and strong currents. Presently, seaweed begins to peak through the disintegrating silt cover. Everyone observes that the time is nearing for macha's sandy beach bottom to reemerge, swept of muck and open for juveniles to attach themselves.

Divers are the first to definitively sight tiny macha in the spring (September to October) of 1999. But from the managers' perspective, the natural process of a bank recuperating from either overexploitation or total loss from a climatic event will require quite a period of abstinence. Of course, this is neither what harvesters believe nor want to hear. Managers are intent on reining in the waders who lacked self-control in their home bay, given current tender environmental conditions. They start by demanding that all San Pedro waders register, despite strays having to be left unquantified.

Meanwhile, the managers are counting on divers to return under their caletas' control like before the flood. Managers want them to revive enthusiasm for conservation within the Management Area. Privately, they're fearful that divers might feel that renewed strict self-regulation would be a waste. Paradoxically, the flood's obliteration of self-managed macha becomes the government's political impetus to reverse gears and speed up Coquimbo's application.

Meant as an enormous morale boost for the caleta divers, the government dubs the richest, most carefully tended area as the first official Management Area in Coquimbo Bay—signed, sealed and delivered in August 1999. Ownership rights are held by Caleta Peñuelas, but the two caletas will share. As the divers triumphantly celebrate their official Management Area and the reemergence of wee macha, the bank is inundated by waders.

Like the managers, divers have less tolerance than before for waders, being vested in protecting the immature, struggling stock. It infuriates the divers to see their tiny macha preyed upon by San Pedro and the strays living along the beach in front of La Serena, despite social sympathy for everyone suffering from the slow economy of Chile's continued recession.

Waders' clear option to migrate is now under such close surveillance that too many waders decide to stay behind in their home bay. Under the logic of their old practice they would have left the macha to recuperate. Instead, waders feel provoked into trying to survive on small macha.

In an effort to respond to the macha predicament—one they were inadvertently prompting by hemming in the waders—the government invests in the university-run nursery area project. The biologists recruit and pay San Pedro waders to find young macha and transplant them to the new Management Area. The aim is to restock this centralized bed, which preflood had been their most productive and guarded.

Scandalizing Tourists

It was the 18th of September—Chile's Independence Day. Members of the La Serena tourism sector are distressed when an intense scallop-transplanting project happens to overlap with the annual holiday week. They don't want to scandalize tourist families who regularly flock in for a spring vacation along Coquimbo Bay's beachfront.

Julio, a dashing family man who runs one of the small beach hotels is in an agitated state. We're sitting under a candy-striped canvas awning flapping in the early morning sea breeze. The hard wet beach spreads out beyond us at low tide. Julio wrings his napkin, repeating over and over that he doesn't know how to hide those San Pedro waders undressing and dressing right in front of his hotel. "And right here at the heart of the bay. What a sight for my guests!"

Oblivious to this concern over dress codes, twenty meters down the beach the waders are carrying on as usual, hurrying to change out of clingy wet clothes. Their horses and ox carts are waiting impatiently to haul away the catch. Julio shudders, pointing at the backside of a particular horse relieving itself right there on the sand, and says, "Look how they're leaving piles of manure everywhere, inappropriate for a fine tourist spot."

A week later, I return to see how Julio is doing with the waders. He confides that their tourism chamber group took action and has been lobbying the powers in charge of local fisheries at the National Fisheries Service and the Coastal Use Advisory Board. Julio says, "We've explained how rude the fishermen are, and how critical it is to reserve the first 200 meters running out from the beach for our tourists."

Instead of heeding this plea from the tourist sector, fisheries managers and university biologists sway the Coastal Use Advisory Board in the other direction. In light of the El Niño flood, the board backs the nursery project to revive macha within the Management Area, which lies completely inside the first 200 meters coveted by tourism.

The resettlement project turns into a fiasco, waders selling off most of the juveniles en route to the Management Area. Waders have sabotaged the nursery project partially for immediate gain, and partially because they harbor resentment toward the Management Area, which they fear will exclude them. A longtime wader, Monchi, says, "Why help resettle the tiny macha to build up a centralized bed where we'll never be welcome?"

After the project prematurely runs its course, Cariaga with a few other San Pedro leaders persuade their people to take careful stock or risk banishment. They finally decide on their own quota in the hope for tolerance inside the Management Area. With this polished patina of good practice, caleta waders continue harvesting the fragile beds of new recruits, but in a more orderly fashion.

Despite the divers' qualms, waders had not already pilfered the entire stock. However, leaders of Caleta Coquimbo and Peñuelas, Cortez and Rene, want these waders stopped. To this end, they ask local marine enforcement to kick out the waders. But, the officials are unwilling to ban the seemingly orderly waders adhering to quotas. One Peñuelas diver, Mateo, describes what happens next, saying, "We take action into our own hands, leaping into our boats. Devil be damned for all our previous self-restraint! Why let waders take it all?"

Here is a classic tragedy, where the fear of others harvesting drives everyone to grab as much as they can. What results is a free-for-all of divers versus waders in the end of 1999 (summer), where there had been, and will be again, a fine macha bank. The antithesis of management practice, waders and divers are unwittingly sabotaging the bay's recovery, competing over immature juveniles clinging to the sand. Nature ends the contest, the bed only able to sustain this two-sided, double-pincer harvesting for a week before there's nary a macha to be found. Both camps withdraw in a huff. Any alliances groomed among divers, waders, and biologists have collapsed.

Before, Cortez and Rene had persuaded their divers to cooperatively guard the Management Area macha stock while abiding the presence of waders. Battling over the last tiny recruit—the destructive human fillip on top of the flood—overrides previous toleration. Divers' relations with the waders falter to the point where they heed the biologists, instead of ignoring waders, during this delicate period of recuperation. They shift emotional grounds, making an acute social swing against waders. Feeling less torn between their two groups' shared social concepts of freedom to harvest and a sense of solidarity among fishermen, divers are rethinking and challenging the historic and humanitarian claims that had allowed them to coexist with waders. Divers are seeing waders as personifying why the macha—the fruit of their management efforts—cannot recover posthaste. They throw themselves behind the biologists to urge the termination of wading, waving their new "legal responsibility" to limit entry to the Management Area for the biological good of the area.

I ask one of my Coquimbo biologist friends, Benji, about having focused on Management Areas for divers, and concomitant reticence to help the waders. He explains the dynamics, saying, "We always understood that juvenile macha grow up inside the waders' zone. We knew waders were problematic, judging them biologically speaking, so we've been much less forgiving toward them. Divers, who share waders' sense of social welfare, had to struggle to break that. It took this *fracaso*—fracas. That's the sunny side."

At Peñuelas beach, I ask Davi, one of the more vocal young divers temporarily longlining, how he felt. He's tickling and tossing his toddler son beside his wife, who is baiting the myriad hooks. He doesn't wallow in self-management's past fruits being buried, which would have surprised the biologists and managers. Instead, he sees how the biologists have been worried how the disaster could undermine conservation morale, thus pushing extra hard for a timely Management Area reward, bringing marine tenure and harvesting rights at last.

Candidly Davi says, "A lot of us didn't regret cutting back the number of days we dived, or our boat's daily quota, because the prices kept rising and compensating us for our smaller haul. Cortez and Rene had explained the benefits of reducing operating costs and labor time to earn the same income. The predictions came true. We were profiting, not tightening our belts."

Davi continues, "I like playing the chance game with the sea. You've seen La Serena's fancy casino full of gamblers; well, we're gambling every day and take what comes as chance. Who knows what's next. There is never the same old El Niño twice. Even that flood was unique. Have you been up to see the dam they're completing?" I tell him that I've driven up there. Far above Serena, at one point the road narrows and snakes around, with a pullover to stop and look out across a gorge and the massive crescent dike almost cementing Rio Elqui's natural flow.

Soon, the macha seem to have disappeared or been depleted almost everywhere in Chile. By 2000, macha definitely are invisible inside Coquimbo Bay. Of course, the local price responds by soaring. The few buyers left on the beach pay hard cash to anyone bringing in whatever they can scrounge from neighboring bays. Like the divers scrambling to make money elsewhere, the three main macha buyers have shifted to other markets or jobs. One buys local fish; another runs a bar; and the third is the one who'd initially gathered up the mass of Peñuelas divers for a stint down in cold Quellon.

Divers classically negotiate individual sale arrangements—*compromisos*—for debts and services with different buyers. Caleta leaders realize the advantages of creating a unified macha marketing strategy, something they hadn't done before the collapse. However, the two caletas are not of one mind on how to proceed. On the one hand, Caleta Coquimbo's Cortez says, "It would be best to evolve beyond Management Area regulatory measures into a type of consolidated marketing group."

On the other hand, at his monthly caleta meeting for Peñuelas's divers treading water, Rene stresses that they have a different ideal, preferring to gamble with fluctuations, in contrast to a fixed contract. "We'd like to auction the day's catch as a group all together on the beach, establishing the day's price for anyone who wishes to sell this way," says Rene. "And then, whoever still wants to can sell their macha individually to various buyers."

Dreaming ahead to future macha, one diver is nudged by his buddies and says, "I love days when we win the big price, like when all the buyers are fighting over Coquimbo Bay macha because the other places were too stormy. I don't want to lose that chance."

There's lots of nodding in agreement as he sits back down. Globally, artisanal fishermen often enjoy the excitement of the hunt as well as the thought of the thrill betting on that day's price. Never mind that in reality, most fishermen fall into a security web of financial arrangements with a buyer—including credit, debts, or loans—in exchange for the promised catch.

René works longlining hake along with half the Peñuelas divers—about one hundred strong. Most of the others are making ends meet by doing odd jobs such as urban construction, roadwork, driving, or baiting for the caleta's other fleets. Cold Quellon still holds a score of their migrants. Very few can start a small business like the little restaurant Rene has opened with his wife.

Caleta Coquimbo's macha divers only number a quarter of Peñuelas's—about fifty. The majority have delved into deep diving as alternative work. This is the real distinction from Peñuelas divers. Their others' strategies paint a familiar pattern. Some are longlining, and one crews on his brother's purse seine boat. Another group labor in nearby Tongoy Bay's scallop-processing plant. Several found jobs in highway construction; one drives a collective taxi, and two drive trucks. A couple of divers have migrated north to dive octopus around Iquique.

Condo versus Caleta: Class-Clashing Lifestyles

The specter of increased urban development looms along the bay. In between Peñuelas and San Pedro there's already a long stretch of high-rise condos mostly servicing La Serena's summer tourist industry. Tourist-generated sewage and garbage have caused terrible summertime pollution since La Serena became a destination for thousands of Argentineans and Chileans.

When speaking of tourism, caleta members reverse the common public rhetoric that points to fishermen as key polluters, déclassé, boats and gear strewn around, dead seaweed in heaps, and even indecent exposure. In an interpretation focusing on the visual realm, instead of increased sewage, Davi says, "Tourist rubbish floats in, and we kick it aside to pull up our boats. We've had to jettison their bottle tops mixed in with the usual rubble of rocks and shells along the bottom."

Fishermen, with working-class sensibilities, classically clash with tourist developers or professionals who move in as second-home owners and retirees. In the Florida Keys (Meltzoff 1997), zoning battles raged over offending aesthetics and historic rights, as well as rising taxes for social services and the coveted ocean-view real estate.

The more northern bay shore out by San Pedro has yielded macha in the past. No Management Area has been planned here, although the dive caletas did discuss the possibility of laying claim to the area. It is unlikely that government will grant a Management Area when there's already conflict between biologists and waders, let alone tourist interests seeking to dominate the bay, as in their failed attempt to claim the Management Area's first 200 meters.

A Spanish consortium recently bought up government rights to construct fancy hotels and condos around San Pedro. This development originally planned to displace San Pedro's population. The La Serena municipality has stepped in to organize a compromise for San Pedro, shifting construction sites slightly to the north. In an even more major move for the community, La Serena has asked the government to fund a caleta-wide training project to transform the caleta's identity from fisheries to gastronomic center. In this scheme, waders would service the tourists and urbanites, their caleta becoming a place of cute cafes and restaurants, serving fresh seafood caught by the two remaining caletas. Depending on how ownership and profits were shared within the caleta, urban development could either dispossess or bless the wading families of San Pedro. But to get off the ground, all aspects require an end to recession.

CONCLUSIONS

Fishermen expect nature to be unpredictable, but Rio Elqui flooding into Coquimbo Bay shocked everyone. It was an unprecedented circumstance. An irrigation dam on the Rio Elqui—planned long before the 1997–1998 event and already under construction—will prevent this unforeseen scenario from ever reoccurring. Fishermen accept such seesaws where they can win or lose. They rejoiced when the great 1982–1983 El Niño event in Coquimbo Bay generated great harvests of macha and loco—the nation's prized abalone-like shellfish. Then they suffered the 1997–1998 event's terrible flood. "Who knows what can come along?" says one diver, Alberto, wiping a wisp of gray hair out of his eye.

Nobody predicted Rio Elqui's surging debris into Coquimbo Bay, even after the event's expected heavy rains commenced. After the flood's devastation of their macha, Coquimbo Bay divers and waders with their leaders sift through past press reports to see if they should have known to do something. Alberto again sums up the consensus, "The press is good at reporting dreadful news of what happened and at least a version of what's going on now, but not what will come. Who can have that crystal ball? Rains came down in a deluge too swiftly to rescue a harvest."

After total macha death by suffocation from sediment, biologists working with the caletas to develop Management Areas became worried whether leaders will be able to continue persuading their members to keep up the strict self-management. They wonder out loud if once macha grow back, the caletas will resume conserving for the future. The biologists do not share the fishermen's perspective that takes arbitrary risks of nature in stride. It's the human-caused problems of government blocks that undermine a caleta's willingness to stay committed to formal Management Areas.

Caleta divers in alliance with the biologists have been entangled in a decade-long political battle to prove their ability to own and control a Management Area. Basically, under leadership and peer pressure, they'd obeyed their own stringent limits even as the process dragged. They were angling for Management Areas, proving their ability to control macha to Chilean officials who were dangling marine tenure rights over fine beds in front of them.

Caletas and biologist allies were equally unaware of how the government would drag the Management Area process out for years throughout Chile. Numerous mapped and decreed Management Areas were being stalled at the final critical step of becoming official, and thus harvestable. The most popular theory circulating the caleta for why the Chilean government is dragging out the process, is as Cortez says, "The government has mixed feelings about trusting us fishermen with ownership of marine areas. Probably it's due to their fears of us controlling places in front of rich coastal developers. You know how angry owners of the tourist spots get when we're working on the beaches by them."

The Chilean government's overarching fear has been of losing control of marine tenure by ceding rights to caletas. The Management Area concept initially might have been less contentious among officials and politicians if the status conferred only fishing rights instead of marine tenure's control over a specific underwater realm. Officials mistrust fishermen's ability to cut back harvesting, pointing to their lack of willpower and the comprehension of biology needed for self-management. In post-flood Coquimbo Bay, however, the government finally grants divers formal Management Areas where they'd been carefully caring for the beds. The award of one of the first Management Areas in Chile came as a boon to boost morale and keep them on track, a rare gift to them and their biologist allies to promote continued self-control despite total loss from nature.

With an official Management Area to reap once macha return to the bay, divers dream of harvesting again, though there's nary a fat macha to be had yet. Without a harvestable stock, they cannot afford to patrol the maturing juveniles in the intertidal zone. For the divers, it feels like a catch-22 because they can only afford to post guard boats where they are actively harvesting, not on a vacant bed.

At least the dedicated university biologists work for free, but the national Management Areas system demands other fees from caletas. The Coquimbo Bay caletas might be forced to jettison the Management Area, if it were to drag on too long without yielding a profit. Thus, these strapped times cry for special government consideration for funding assistance, or at least a plan to defer payment of Management Area costs. No safety net exists.

Just as life's first glimmer becomes visible on the bay bottom, biologists' worst fears manifest. Divers, waving off previous conservation, make a mad dash in a race with waders who have begun collecting the fragile stock. They fall prey to the tragedy of the commons, both groups grabbing as many baby macha as they can from the bank. Divers and waders competing against each other from either side of the surf zone, macha could only endure a week until none remained. Coexistence between waders and divers in Coquimbo Bay has sunk to a nadir. Except for the week-long fracas, divers fall back in line under caleta leadership with the biologist allies. With the Management Area conservation process that has finally awarded them marine rights, they are more motivated to turn their backs on former shared sensibilities to let anyone fish freely when in need. These historic humanitarian rights that everyone felt waders deserved to coexist in the bay pre-flood are conveniently usurped in the divers' minds by the marine biologists' logic.

Waders up until this time had been relatively uncontrolled, but now government stepped in. San Pedro waders, with no hope for a Management Area to call their own, started facing far more demands to register, and stricter enforcement of regulations in the water than ever before. In the eyes of the university biologists, the waders should have been curtailed sooner; waders unwittingly collect mostly juveniles because their shallow zone inside the surf is exactly where young macha mature.

Divers and waders seek alternative income until their bay's macha can recuperate. They respond to this grave situation, underscoring decision making for survival, by embracing social and natural opportunities to support themselves: migrate when necessary to allow recuperation; harvest when possible; and travel to profitable work in booms. The 1991 Fisheries Law blocking migration harms them, so it is overlooked. They play the dire straits card of temporary legal and social leniency that tolerates outside fishermen, and even the urban poor, as a social safety net.

Both waders and divers have temporarily dispersed until Coquimbo's beds resuscitate, a few landing jobs from construction to driving. The Coquimbo longliner fleet took in some divers to keep them afloat for a while, and managers looked the other way. But waders lack the social access and skills to be able to switch to other fishery groups within Caleta Coquimbo. Divers were also recruited by buyers to migrate south to dive other lucrative shellfish. Waders from Coquimbo Bay physically pushed waders in Tongoy Bay to accept them for a bit.

The lethal coat that smothered all macha, precipitated deeper conflicts straining social cooperation among fishing people. Once macha can be harvested again and it's not a moot point who can harvest, divers might offer San Pedro waders a foot back in Management Area waters. But, divers with their biologist allies have become more urgent and strident in protecting macha from waders. One wonders if divers who now own Management Areas will return to preflood tolerance, or hold the wand of marine conservation up against their rivals whose shallow harvest zone is home to the juvenile macha.

EPILOGUE: BALANCING CONTROL AND LACK OF CONTROL

Nobody seems to be confronting another grave problem physically damaging the macha's recovery. Caleta Coquimbo's shrimp and prawn trawlers drag their nets over the Management Area daily en route in and out of the bay, despite a law banning trawling anywhere inside the bay. Thus, trawlers never leave the Management Area bottom in peace to rest and recuperate. In response to divers' pleas, those in charge of bay fisheries surveillance—National Fisheries Service and Navy—say they've never been able to enforce this ban. Nor have Caleta Coquimbo leaders curbed their powerful trawler group. This dilemma seems destined to continue unless the political and economic forces at play are investigated and negotiated by those with an interest in the Management Area.

Divers and waders feel macha will return to them and stabilize relatively soon. But managers define a stable stock as taking far more time. This is time that divers with a Management Area cannot afford, either in terms of money to guard and run the Management Area or to feed their families with a familiar and beloved livelihood. Social costs of guarding the Management Area against waders would be visible. The waders, as well as the urban poor who wade, attract negative tourism attention through their beach practices. Waders could be integrated into the macha fishery in various capacities if there can be cooperation among the dive and wader caletas. If divers continue their Management Area management in Coquimbo Bay, one wonders if they will relent toward waders and find mutually acceptable ways to incorporate them.

If the recession comes to an end, then the external plan for San Pedro tourist development will pop up around the caleta, converting their place into a restaurant zone. Otherwise, waders could be assisted by alternative government-funded projects, yet ones of their own choosing, unlike growing flowers. But randomly blocking the urban poor who wade to subsist, and stay outside any formal organization, is problematic. It grates against an old European and perhaps Latino concept of social fairness with the economic imperative to allow the poor to glean the fields or flats to make ends meet. Most citizens view it as socially good to grant these situations.

A core question remains among the university biologists who have worked on initial phases of the Management Area studies throughout Chile. Will both government and caletas keep faith in and continue supporting the Management Areas system with its costs and stringent harvesting quotas, despite periodic radical environmental change?

Coquimbo divers and waders see the inevitability of nature changing, and they surrender to its lack of predictability and their lack of control. The rational Descartes scientific Management Area system may be able to maintain divers on a self-management track, working together with biologist allies and the managers as a means to harness biological logic and the marine tenure control over the Management Areas. The young Management Areas system will need to be reformed, and government promises will need to be kept more readily.

Herein lies a kind of philosophical balance. On the one hand, the parties need to accept that even with science, one can't predict everything. On the other hand, they have to agree to adopt rational management techniques and biological logic—despite the limitations of such things. I've reached the conclusion: "Be logical with visible evidence and scientific while understanding nothing provides guarantees." It's provocative, and reflects an acceptance of impermanence and ultimately not being in control.

LETTER: RED FLAMINGO WINGS OVER SCALLOP BOOMS IN PARACAS, PERU

Hi Rico and Gletys,

Desertscapes morph and continue revealing fabulous coastlines of barren cliffs and tiny fishing ports as I head north of Arica's Chilean border into Peru's loco smuggling port. Excuse me, *chanque*, the common name on this side of the border. Though I will return to northern Chile, it is important to widen sites into Peru, to compare firsthand fishermen's survival strategies under conditions of climate and political change. And who could make a better fieldwork hand but your Peruvian soul?

Spiraling over the Nasca Lines in a piper plane, the pilot swoops to give us better views of the giant monkey and spider. Climbing down into Nasca's intricate underground aqueducts also brings up the mystery and intelligence of the pre-Columbian world that we are superimposed on. Nearing the center of Peru's south coast, I reach Ica's enormous ancient inland seafloor now exposed and fossil strewn, from skeletal whales to ancient sea lions. Beyond this, the feeder road dips west through the sands into today's elite Paracas seaside beyond Pisco. You will arrive soon!

It will be wonderful sharing our little Pisco casita, wandering at dusk among women selling fish in front of the Pisco caleta as you name them for Gletys and me, and select the right ones for one of your delicious fish fry dinners. The world of the fishermen and their salt roads spinning deep into the Paracas National Reserve are left unexplored until you join me. I have decided to focus on Pisco-Paracas since twice scallop banks have thickly carpeted the great Bahía Independencía—BI, Independence Bay—due to El Niño events when hoards flooded the area. Some local divers described what they watched during the infamous battle of Laguna Grande. Here's the fight story from several fishermen who were there!

Red flamingo wings overhead against blue sky on the 4th of July, 1998, and fifteen boats of agitated local Paracas divers have aligned themselves to block the *boca*—the mouth of the passage into the Laguna Grande cove where the wharf lies at the bay's northern sequestered tip. Armed with heaps of rounded beach stones and sticks, these locals are determined to repel all foráneos—foreigners, migrant fishermen from other ports of Peru—trying to break through the flotilla to reach the wharf offloading point where scallop buyers with trucks await the precious cargo. *Foráneos* have sparked the battle by raiding locals' scallop grow-out corrals. Because these sites aren't government sanctioned, locals are defending what they feel is their turf themselves.

By late afternoon's long shadows, the usual time to return to shore to sell, at least 130 boats joust at the *boca*, ramming and slamming. Sore feelings soar, and a number of heads are split. Luckily, nobody's killed. The foráneos—outsiders or foreigners—make it through to the wharf, but a few of the Laguna Grande divers purportedly sabotage them by pouring fuel on top of sacks of scallops loaded in a truck to go off to the market. The pitched battle still rages over marine rights within the reserve.

The story for Paracas unfolds as follows, focusing on how the periodic mega El Niño events cause market booms in valuable species, including scallops, here as in Chile. So many get hopeful when the media or websites expect/suspect another coming event. They dream of another wild market boom. The first scallop boom followed the colossal 1982–1983 El Niño event, resulting in an unregulated free-for-all as people poured into the Pisco-Paracas Region from up and down the Peruvian coast, similar to boom migrations up and down the coast of Chile. The second huge 1997–1998 El Niño event triggered a boom.

In the second boom, hookah hose diving took the lives of countless reckless boom-time scallop cowboys who would bounce up and down repeatedly, never properly decompressing in the rush for ever-more money. The death rates didn't approach those of the first boom when corpses still in their wetsuits lay on the piers, the frenetic action continuing around them until it grew too dark to dive.

Out in the quiet chill of early morning sea in Laguna Grande, Tío Rico and Víctor prepare to check their scallop lines. They've taken me along, and Víctor says, "We're really one of the few boats left working on scallop aquaculture and diving outside a boom." I know they're part of the handful hell-bent on aquaculture from speaking with the local scallop exporter who still backs them. I watch from the bow as they don woolly old sweaters, then the outer layer of handmade wetsuits. A light breeze springs up and ruffles the surface as the two men carefully check air lines in preparation for descent. These two take no chances, however, having all the time on their hands and no competition driving them. They take turns running the compressor and spotting each other in the water as they go down to inspect the lines, check out growth rates, and look for signs of seeding.

Tío Rico tells me, "Our grandfathers in the 1950s and 1960s stayed inside the boat, rocking their incredibly long rake handles back and forth to scour the deep-bottom scallop banks, at depths we still can't harvest by air hose." I describe how I'd seen men in Illa de Arousa, Galicia, using similar long rakes for shellfish (Meltzoff 1995). We talk about how these rakes were before the advent of the more efficient air hose technology, before the shellfish export markets, before the booms.

I think this time we will only see coastal people here wishing for the next great El Niño. Wishing we could return whenever the next great unregulated free-for-all occurs, but so far, no more massive events despite all the local prayers.

Happy to have you both here soon to share more of your Peru. Times are always interesting as the soap opera of life unfolds in unforeseen ways. I hope you'll take me to that Lima museum with the hairless dogs running around the grand grounds and transforming the pre-Columbian clay pottery dogs into ones we can pat!

Warm hugs to Mama Hopie in Lima, too,

Sarah

Chapter Five

Puerto Aldea

Caleta Life in Navy War-Game Waters

GAMES IN THE GALES

On one of Tongoy Bay's perpetually windy spring days, Chile's triannual military war games are in force. Navy tanks are barreling along outside the high mesh fence surrounding Puerto Aldea, perched on a bluff above the bay's southern rim (figure 5.1). The fence separates this fishing settlement from desolate, steep, dry hills running farther south down the scarcely populated coast. I'm sitting at the breakfast table with my hosts, preschool Pepe and his mother, Susana, who don't seem to notice the violent vibrating of their house walls from the gale. Pepe laughs when I jump at the added shake from exploding missiles. They are accustomed to this force too. A man crosses in front of the kitchen window, waving and leaning harder into the wind to make progress. He is Julio, a Coquimbo man who owns the shack next door and comes here to buy *lapa*, a nationally popular shellfish. He also collects and sells the exploded shell casings.

Susana says, "Julio is lucky. He almost got blown up when his discovery turned out to be a live grenade that went off." Pepe runs outside to ride his bike and I go to watch the exercises. Tanks rumble through the entrance gate by the little church where the women inside are holding daily catechism, "Ave Maria, Ave Maria." The tanks send clouds of dust sweeping across the village. Nothing is paved inside the village fence or along the tracks leading in and out of the caleta. Everyone is supposed to ask the Navy representative for the key to venture beyond the gate, even the children.

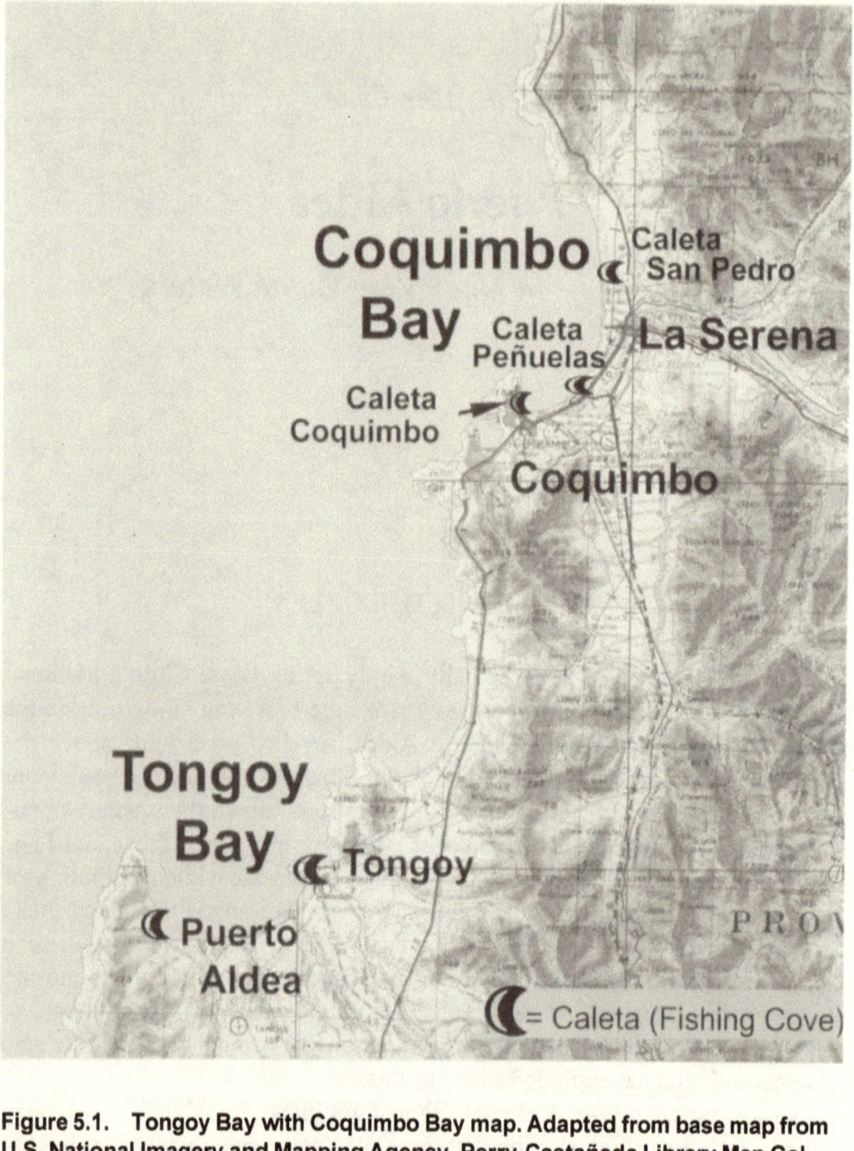

Figure 5.1. Tongoy Bay with Coquimbo Bay map. Adapted from base map from U.S. National Imagery and Mapping Agency, Perry-Castañeda Library Map Collection, University of Texas, Austin.

ETHNOGRAPHIC STORY AND EMERGENCE OF GLOBAL THEMES

In the major 1997–1998 El Niño, heavy rains in June and July 1997 are compounded by exceptional summertime snow melts in the mountains. The "flowering desert" and the enormous recruitment and rapid growth of scallops in Tongoy Bay are happy events. Storm surges also cast scallops up onto the sands of Tongoy Bay, and the chaotic scenes of people scrambling to salvage scallops make the press, fitting the classic national and international popular media images of death and destruction in the wake of El Niño events.

This chapter, like the previous one, pivots around the 1997–1998 El Niño event when Region IV's two great shellfish bays of Coquimbo and Tongoy experienced diverse consequences from the event. Although it's an anomaly for such bays not to thrive with events' warmer waters, Coquimbo Bay suffered a surprise flood that suffocated all shellfish. Until this point, both bays' caletas had been imposing self-management to gain marine tenure via Chile's Management Areas system. However, the government dragged out the process. The event radically changed the availability of shellfish in opposite ways in the bays of Tongoy and Coquimbo, altering the intricate dance over the resources among the players—stakeholders, in development parlance.

Players include divers and waders; caleta leaders and their biologist allies; government managers of the Management Areas; the Navy, which holds tenure over half of Tongoy Bay; Tongoy's scallop aquaculture industry, which holds sway over the other half of Tongoy Bay; shellfish buyers and exporters; and coastal tourism developers along both bays and filling in between. Watch how the players interact within a bay, and between bays, let alone at national and international levels.

Key themes include the role of local leaders and a pattern of who is drawn into these roles, and the conflict between biological management and a deeply ingrained societal consciousness that makes exceptions for those in need.

Compare these two overlapping cases with Paracas, Peru (Meltzoff et al. 2005), where fisheries people are yearning for the next powerful El Niño event to lead them into an absolute riot of scallops in their boom-to-bust cycle. In the Peruvian parallel to the Chilean Management Areas process, fisheries associations seek aquaculture leases from the government in an excruciatingly convoluted, drawn-out political process. Aquaculture livelihood is meant to promote stability between event-triggered booms.

Tongoy Bay

Tongoy Bay's northern half is noted for scallop mariculture; production operations are based in the small town of Tongoy. The bay's southern half is a Chilean Navy preserve, the Navy having taken it over early in the Pinochet era (1975–1979) to hold military games. On the far side, there is one lone fishing caleta, Puerto Aldea. A vast, uninhabited, hard-sand beach stretches for miles, separating the caleta from the town.

Puerto Aldea's divers harvest wild scallops and other shellfish; the women lead in collecting nori and other seaweeds. Their caleta has been in a protracted government Management Area process. The caleta is growing increasingly exasperated with unfulfilled government promises to grant them the Management Area. This is part of an overarching national pattern that Professor Wolfgang Stotz (long-standing head of the Ecology and Resource Management Group, Marine Biology Department, Universidad Católica del Norte, Coquimbo) has been critiquing and working to improve (Meltzoff et al. 2002). Without official Management Area status, harvesting is still forbidden in areas where the caleta has been self-managing to prove they can control their proposed Management Area. When the warm waters from the 1997–1998 El Niño trigger scallop spawning, loss of most natural predators, and rapid growth, the plethora of scallops tempts the divers. Self-righteously, they revert to unbridled scallop harvesting. Here is where my story unfolds.

At dawn, Raul and I set out from Universidad Católica del Norte in the Ecology and Resource Management Group's pickup. Professor Wolfgang Stotz, my new friend and colleague, has taken me under his wing and asked Raul to introduce me to people in Puerto Aldea where they've been monitoring scallops and carrying out the intensive biological studies required by the Management Area process. Raul is to set me up with a family with whom I can spend a few nights. The university nestles at the base of a bay's peninsular cliff. Sun coming up over the hills reflects off cold water, illuminating the wharf with bobbing research dinghies and a clutch of rescued injured pelicans adopted by students.

We leave behind the small port of Coquimbo next to the elegant old city of La Serena that rises above the core of Coquimbo Bay and serves as regional headquarters. Because of the intense El Niño rains, wild flowers are in mad profusion along this normally brown route.

Along the sixty-eight kilometers of narrow coastal highway running southeast of Coquimbo, real estate developers are filling in the rugged shoreline between the two bays, decorating desertscapes with upscale adobe enclaves. Idyllic names arch over the entrances. We reach the town of Tongoy where a long-established exclusive settlement of second homes wends along the peninsula rim with gorgeous bay views. Tongoy is sited on such rich protected waters, aquaculture interests have developed a highly organized scallop industry. I count twenty-seven scallop aquaculture companies in Tongoy. At the entrance to town at the base of the peninsula, row upon row of workers' housing line the back streets, and there's a burgeoning school. Migrants from the capital of Santiago to Coquimbo work here, especially in the summer season, when waters are warmer and scallop harvesting and processing peak.

Raul pulls up in front of Navy headquarters and disappears inside while I wait in the jeep. One must request and register all visits to Puerto Aldea.

Then we sit at little tables set out on the sands, savoring scallops for lunch while watching the tide recede until there is enough flat, exposed beach for the jeep to round the bay's crescent. Shore birds dart and scatter in front of our wheels. Stopping here and there, we get out to toss useful flotsam—like fishing floats broken loose in storms—into the back. And because it's low tide, we can get out to watch different groups of people collecting seaweed or wading back out of the shallows with *macha* bulging in net bags hanging from their waists (figure 5.2). Wetsuits peeled off, some waders are sitting on the hard sand, sorting through their haul to sell to the truck buyers who've also driven from Tongoy along the long beach (figure 5.3). Finally as we're nearing the far side, we sight Puerto Aldea perched on the bluffs.

Ascending in low gear to the Navy guardhouse, Raul hands over our Tongoy permit and a guard lets us enter the gated settlement. The Navy keeps strict control over who enters their war-games territory. Only later will I learn how the Navy is ultimately blocking the caleta's Management Area because it's in their war-game waters. Control is the game, loss of control the issue. Raul drives me up through the village and around, and loops down to the wharf. A few open dories rest upside down on the ground being brightly painted by men he recognizes. He calls out names and we all shake hands and hug in the standard warmhearted Latino salutation and parting gestures. I come bearing introductions and greetings to caleta members from their co-op leader, Nelson, who I'd already begun interviewing at his Coquimbo-based office. He has already told me a bit of his life and work history, and briefed me about the need for clandestine scallop harvesting since their Management Area remains unofficial.

There happens to be a big Navy vessel standing offshore. Jairo, one of the guys painting boat bottoms, waves his brush toward the ship saying, "Warship. It ducked into Tongoy because the war game is about to begin. The sailors might come onshore, and then we'll hold a barbecue together, lots of music and dancing."

I see how the caleta associates the Navy with breaking the routines of isolation.

Visiting the grade school the next day, instead of houses or jets, I see the kids have drawn warships, taped up sailing across the walls. The Navy has become part of childhood imagination in Puerto Aldea. Erica, a young single teacher from Serena says, "During peaceful maneuvers, a crew may ferry the school kids out to climb around the ship and have cookies and ice cream. It's fun!"

By slanting late light, Raul drops me and my duffle up at Susana's. She is letting me a bedroom in their home with her little son and husband. A lean, tall, tawny woman, she wipes strands of hair out of her eyes and shelters me inside from the winds. The jeep will return in some days. Susana sits me down at the kitchen table with a warm bowl of fish soup and tea as Martin

Figure 5.2. Tongoy Bay macha wader with his mesh collection bags. Courtesy of the author.

Figure 5.3. Tongoy Bay waders sorting through their haul of macha to sell to the truck buyers. Courtesy of the author.

crayons blue trees. We begin getting to know each other. Born into a family of herders, not fishermen, Susana has lived around this caleta her whole life. She lays out the story of Puerto Aldea. It's corroborated by the other long-time inhabitants during later visits.

Prior to 1975, Puerto Aldea had been a tiny place of migrant shepherd families, like Susana's. Then as a result of the Navy takeover, when land and grazing rights became rent free, a number of rural families from surrounding areas settled in. Concurrently, caleta growth between 1975 and 1986 was driven by men arriving to ride the wave of the boom for costly *loco*—an abalone-type shellfish that Chileans savor and prize. Because of this nation-wide market demand for loco, Puerto Aldea's houses and boats doubled in number. Fishing became the focus, and herding—more relegated to women—became secondary.

Settlement swelled again when Chile enacted the 1991 Fisheries Law (*Ley de Pesca*). Its ambition was to end the highly migratory lifestyle of artisanal fishermen who expediently moved around in pursuit of resources and market booms. At the time the law was drafted, managers were distraught over the rampant ongoing loco wars. They blamed the migratory nature of fishermen and felt that settling and registering fishermen would lead to orderly management.

The caleta's Mayor Reinaldo (*Alcalde de Mar*) comes from the Veliz family, who were the original settlers, and thus the main ones to lose their land rights to the Chilean Navy. I appreciate the touch of ironic justice that he now serves as the liaison with the Navy. He proudly shows me the big model Navy ships he keeps in his yard overlooking the caleta boats far below (figure 5.4). As mayor, Reinaldo has been custodian of the Navy's key, and acted as naval gate guard up until a recent posting of Navy personnel to this outpost. Mayor Reinaldo owns and runs the little pool hall, which is the font of local beer. This is a perk to him as well as his customers, because although alcohol is technically illegal on Navy land, the Navy looks the other way on such a minor issue in the oceanic scheme of things.

It's a coincidence, like most events experienced during fieldwork, that I happen to arrive during this week's military exercises, and I feel lucky to observe this aspect of local life. The Navy has ordered the caleta to keep all boats at the wharf. Yet each day I watch several sally forth despite the physical and legal risks. The Navy has also bidden everyone to round up their flocks of goats grazing the hillsides. Yet, many families fail to keep their animals in check. One old shepherdess bluntly says, "It's too much work to

Figure 5.4. View of Tongoy Bay from the caleta mayor's prized model Navy ship. Courtesy of the author.

herd my goats inside the caleta, and to have to build a pen and supply feed when they're content grazing out there by themselves. Chances of getting blown up aren't great."

The physical danger from tank shelling is clear for people, though, who do comply on land. They may risk their animals, but this week nobody dares walk the bay's rocky southern coastline to harvest *huiro*—their best seaweed product.

In front of Susana's kitchen door, I can't help but notice the large mound of dried seaweed. I ask her about it. "My husband buys it all from our neighbors, loads it into that rusty red truck and sells it in Tongoy. Anyone with balance walks along the rocks out that way," says Susana, pointing toward a distant lighthouse, at the end of their peninsula. I'm wishing it were safe to hike out there myself, but the military booming really is intimidating. Susana turns back inside and I follow.

"If someone's not busy on the dive boats, they can still make cash from different kinds of seaweed. Huiro is the one with the golden globe for a head. Just pull it out of the water and carry it home, then cut off that ball and spread the leaves to dry for a couple of days. You can sell it for a nice price, or boil it up in soups," she says, dipping her wide, wooden spoon into the broth simmering on the stove and lifting up limp strips of huiro adding flavors to our lunch.

"What about the old couple I saw collecting seaweed washing up on the beach when we were driving across from Tongoy?" I ask.

"Oh, they're gathering seaweeds that float up onto shore, like *pelillo* [hairy] because it looks like a mop of hair. We dry and sell them, too, but for much less money. If they're not too weak, men dive shellfish from the boats, leaving seaweed to us women," says Susana.

Caleta divers legally bring up piure, taca, jaiva, and chocha, but they may also sneak loco, a national pastime among fishermen with access to this favorite Chilean and Peruvian delicacy. Loco ranks far above all other shellfish, and here in Region IV is the heart of loco harvesting. Fido, Susana's husband, comes in with the wind, taking off his woolen coat, angular nose, head held high, his long black hair blown. Susana asks him about what she's forgotten to mention that the caleta harvests—chocha.

"*Chocha*—a type of shellfish—are particularly plentiful in Puerto Aldea. We're one of the few places in all Chile where they like to grow," says Fido. "Usually, quantities are too slim to justify export buyers' time. But once around 1986, during the time of the great loco boom, I remember a short export boom in chocha."

"Yet of all our marine products, scallops are the king," says Susana. "Never mind that Management Area managers say it's illegal to harvest them!"

Walking down to the wharf after lunch to see who was coming in, I'm thinking about how wild scallop harvesting is irresistible to divers who see the fat bivalves waiting for them on the bottom. Oscar waves me over. I shout down to him in the boat, asking him what he was diving for and he says, "Scallops, of course! It's our right, our top product. You should see them clustered down there. It would be a waste to ignore nature's gift. Anyway, who's here to stop us?"

As Oscar is peeling off his thin wetsuit, shaking off the wet with a patch of towel, and pulling on a dry sweater, he explains the classic hookah technique, nodding to his onboard compressor that pumps surface air into the long hose now lying coiled at his feet in the bottom of the boat. He climbs onto the wharf, leaving behind his crew, Rubio, to organize the gear. Rubio is an older guy whose nubby sweater seems to double his body bulk. "Rubio is in charge of the compressor and our lives, and stays dry," he says with a wry smile. "We go down, biting on the mouthpiece at the end of the hose, exploring as far as the hose extends. I either stay down until I've collected an area, or the cold gets to me. Rubio knows how to run the compressor and maneuver the boat to follow my air bubbles. He must feed out the hose and look for signals to hoist the net bags once they are full of shellfish."

I ask Oscar about the scallop aquaculture sector on the Tongoy side of the bay and how they interact with it. Oscar says, "Aquaculture is loaded down with big investments and high costs compared to us wild harvesters. In Puerto Aldea, what do I need besides this?" he shrugs at his small boat, his wetsuit skin, mask and fins, lead weights, still in the bottom of the boat by the compressor. His point is well taken, his comparative investment is slim. A dive boat operation is largely sweat equity, minimal fuel and maintenance after the initial investment in the boat and small outboard, compressor, and hoses. Because of this, in the market there is a 1:10 cost-ratio: one wild Puerto Aldea scallop costs one peso to produce, whereas one cultured Tongoy scallop costs ten pesos to produce.

Oscar launches into an emotional topic among the divers: how Tongoy boats sneak across the borders of Puerto Aldea's fishing grounds to steal or purchase their wild scallops, reselling them to the aquaculture processing plants. That this violates the law that prohibits the harvesting of all wild scallops isn't the issue. Rather, in this triangle everyone is complicit. I wonder why the industry seems content with this arrangement.

Rubio has hopped up onto the wharf, too. In a resentful tone, he says, "Look, this is not so good for us. We're having to sell scallops under the table to the industry for a price lower than we could negotiate if the government opened our wild market."

Scallop harvesting is in legal limbo with the Management Area, but Puerto Aldea has the trade in nori seaweed. Nori—a seaweed commonly eaten by the Japanese in sheets wrapped around rice and raw fish—is the caleta's most lucrative legal product. Nori is called *chicorea* in Chile. Chicorea's main growth and harvest is during the summer—October through February being the main season—as well as during particularly sunny winters. The community communally has been harvesting and marketing nori through their caleta ever since 1992, when they started organizing seaweed extraction through a new organization with creative caleta leadership.

Puerto Aldea Gains Caleta Organization and Leader, Nelson Plaza

When Nelson Plaza first set eyes on Puerto Aldea in 1987, he was looking for nori to buy. Nelson, as a Coquimbo man with training in marketing, looks back at the time he first came to Puerto Aldea and says, "I was touring caleta by caleta, looking for this precious seaweed. The people in Puerto Aldea were selling their nori at the lowest prices anywhere. They were operating individually, each one for themselves, without any organization, no caleta association." He shakes his head and continues, "I decided to actually move here. It seemed like a good chance to follow my political ideals by trying to organize this community around a high-priced product. I'd been keeping a low profile with Pinochet during the bloody early years, but now I wanted to help the caleta bring in structure and the benefits of better prices."

A number of idealistic Chileans have been motivated to move into more politically isolated and economically naive caletas to try their hand at organizing fishermen. This is an important pattern of Chilean leadership, where the highly skilled, often searching for political anonymity during the Pinochet period, took refuge in caletas in the hope of engaging their own idealism and energies.

Nelson was born to Coquimbo parents, but in Chile's southern Temuco region. Nelson says, "My parents moved back to Coquimbo when I was still a baby. So I count as being from here. After high school, I got commercial technical training and went into the south for fifteen years, even marrying a woman from there. These were the Pinochet years, and my views were dangerous to air in the center of the country. Finally, I got less skittish and returned to Coquimbo, finding work as a seaweed buyer."

Nelson's local stature, higher level of education and expertise in seaweed marketing, ardent political idealism, and charismatic character won the trust of Puerto Aldea's tiny rural populace. They had nobody else to organize their fisheries, and in 1990 when he initiated their caleta association, seventy caleta members eagerly elected him as leader. Nelson has guided them as the caleta increased in size and services. His drive for community fisheries development has been steered by his skilled lobbying efforts and proposal writing.

"When I moved in, there were only forty-three houses with no infrastructure. A decade later, we have seventy-two houses, and there's a church, primary school, freshwater well, and a modern processing shed by the wharf. The latest improvements for our welfare are the new electricity, and soon a nursery school as well as a fishermen's social casino," says Nelson proudly.

In 1991, the Chilean government began moving toward the implementation of their radically new fisheries Management Area system. The government wanted to encourage local participation in management. Its aim was to curtail artisanal fishermen's rampant overexploitation of marine resources and perhaps sweeten the major 1991 Fisheries Law, which demanded the end of migration.

The Management Area's enticing promise was that local participation in management would be coupled with marine tenure—ownership essentially. Leading up to the Management Areas system, competition between caletas over resources could be fierce. No customary marine tenure system existed; instead a competing sense of historical use rights with a concept that the sea should accommodate everyone.

The first handful of caletas that qualified to start the costly and lengthy process were allied with biologists. Caletas that had been fishing in an open-access system without paying sharp attention to regulations, started to set up management-area self-regulating quota systems. They also started guarding their proposed Management Areas, although the official Management Area maps and political decrees would come at least six years down the road. For most Management Areas, final official status would be conferred years later. But unsuspecting of future delays, hope soared.

In 1991, Nelson Plaza teamed up with Wolfgang Stotz's ecology group at Católica del Norte. Puerto Aldea in tandem with the biologists requested one of the first Management Areas. Within a year, Chile's Community Development Agency, which had never before worked with a caleta, had added Puerto Aldea to its list of field sites, and Wolfgang and his group started Management Area prep work.

Wolfgang says, "The Chilean community development agency sponsored our efforts with the fishermen. In turn, we were determined to teach Puerto Aldea about ecological management so they could have their own Management Area. This type of marine tenure can create a sense of responsibility and guardianship, the hope for conservation!"

Stotz's group also assisted Caletas Coquimbo and Peñuelas in Coquimbo Bay (chapter 4). And you can compare how Caleta Chanavayita outside Iquique (chapter 7) with their biologist allies becomes the first Management Area in Chile through the politics of fisheries.

Wolfgang and Nelson got along well. Nelson wanted to apply ecological management concepts he was learning through the university's Management Areas workshops. The concepts that the university was transmitting to Puerto Aldea to make the Management Areas possible centered on the caleta coordinating both harvests and marketing. The caleta needed to start long-term planning to yield maximum profit and conserve the stocks. The Management Area asks the caleta to organize extraction and marketing; repopulate restricted zones; weed out scallop predators; and develop underwater corrals to protect scallops from occasional but serious storm-surge damage.

Nelson as a strong leader was able to help create and implement this caleta organization, and define membership and insider fishing rights, as well as rules of exclusion. Newcomers, as well as locals who had emigrated and registered in other regions, now had to petition the caleta for permission to fish within the Management Area. "I hoped to start harvest seasons inside the Management Area. We began a caleta program of sequenced species extraction, and collectively negotiated prices so we could earn just as much for smaller takes, our type of conservation," says Nelson.

Sergio, on the university biology team, got so involved that they embraced him as a caleta member. Traveling back and forth between Coquimbo and Tongoy bays, he often spent the night, getting wet and cold doing biological studies and helping teach Management Area workshops into the evening.

Sergio says, "I half lived there and got so wrapped up with the success of the caleta and the people that I joined myself. I'd first gone there just for my thesis research on the fauna associated with a rare local seagrass species thriving in the area. But with the Management Area work, I got caught up in political biology and in helping my friends qualify as one of Chile's first Management Areas."

With Sergio working in situ with Nelson and the members, the university implemented a series of strategies designed to get Puerto Aldea's Management Area officially decreed. Obtaining Management Area status was proving extremely difficult. It required the completion of expensive baseline studies and a complex management plan that could only be drawn up by professional biologists.

Gaining USAID funds to cover their effort, Wolfgang continued to lead Sergio and the others in his group for the next two to three years in working with the local fishermen, until they finally completed baseline research on Puerto Aldea scallops. In achieving this, Wolfgang says, "Puerto Aldea probably ended up undergoing more marine biological studies in its waters than any other Chilean caleta in history, including in Coquimbo Bay where we also worked."

Besides biological studies, Sergio and the others spearheaded the training and education program. Optimism was high. Well-attended workshops included classes in marine biology on life cycles of scallops, as well as on nori, the caleta's other main product. "Lively discussions revolved around the economic benefits of long-term marine management. I also wrote lots of letters to help Nelson locate new nori buyers who'd be willing to top the price being paid by their only buyer, who held a de facto monopoly," explains Wolfgang.

One weekly workshop was created to introduce the concept of Management Area reserve areas to fishermen. Another was a three-month televised course on small business management that Sergio also enrolled in.

"Going weekly to the caleta as a teaching assistant, I started taking the small business course along with ten other caleta members. Half of us took the exam up in La Serena and graduated, but I think five guys were too intimidated to go through this formality, maybe because they'd never gone beyond grade school," says Sergio.

The government quickly stalemated approval of the first set of proposed Management Areas. This demonstrates how policy passing into law never ensures its promised implementation. Officials steer by their particular biases; in Chile, the sheer number of national artisanal fishermen votes (about 60,000) politically motivated the government to start the Management Aareas process to win those votes. But throughout most of the 1990s, at the national level, Management Areas remained as concept and discourse. Never mind that caletas with their allies had begun eagerly acting in good faith.

The national debate focused on how to grant marine tenure and how to trust artisanal fishermen. The government never intended a major restructuring of a coastline already occupied by politically powerful interests in tourism, second homes, aquaculture, and fisheries. Politicians were preoccupied with allocating a minimal marine area to please fishermen while basically maintaining open access to Chile's coastal waters. Officials did draw up the Management Area maps, but decreed that only certain fishing grounds qualified as potential Management Areas. Only the few caletas already fishing in these specific areas could apply for legal Management Area status.

Theoretically, the Management Area would become the property of the caleta after four years of proving good Management Area management by accomplishing objectives laid out in their caleta's management plan. However, the government was arguing over every step, with seemingly unworkable demands. For instance, as the marine environment shifted over time, they kept insisting that the costly baseline studies had to be repeated. The baseline study, besides the absurdity of repetition to record constant changes, implied a virgin coastline instead of the overfished reality.

Sergio, Nelson, and I are out in the wind on Puerto Aldea's wharf, watching boats heading in. Nelson says, "After the whole acceptance process, the caleta must pay them an annual rent for the use of the Management Area. This fee is supposedly going to be based on the size of the Management Area, instead of on what we can harvest and current market prices. It's unfair."

Sergio and the others are nodding agreement. It's frustrating all around given everyone's efforts. From 1995 to 1998, Wolfgang Stotz's ecology group strove continuously with Puerto Aldea to complete the acceptance process. The mutual goal has been to make the Management Area a legal and functional reality so they could start harvesting scallops—their desired reward for self-management.

"We had separate financial success when a Korean nori buyer moved into the area and started buying, providing some alternative income. But the caleta faced the problem of not being allowed to sell scallops inside our own Management Area!" says Sergio.

By 1997, the university had finished a pricey project in Puerto Aldea on managing the caleta's wild scallop stocks, including a scheme that involved scallop hatcheries in Tongoy.

"This time," says Sergio, "we had held workshops in Puerto Aldea over a fifteen-month period. Starting with fifteen caleta members, we ended up with three attending. Only one of them passed. Then, our one graduate left Puerto Aldea and moved to Tongoy to work in aquaculture. He told me before leaving that he just couldn't stomach always having to break the law in order to harvest scallops. Nobody else had qualms ducking the legal dilemma, feeling justified because it was the government that was thwarting our Management Area."

Caleta members lost hope over the government going in circles when they needed to harvest. But Wolfgang and his group continued the Management Area mission. Back in his university office, Wolfgang and I discuss all they'd gone through. He is a tall, spare, bearded, strawberry blond of German Chilean descent, trilingual, and highly respected in international, national, and regional circles as a leading marine scientist in ecology and management efforts.

Looking calmly out the window through high pines to the bay, Wolfgang says, "We were wracking our brains what to do, watching the caleta quit. I tried going to the very top, to the sub-secretary of fisheries—the government's head of fisheries—petitioning them to create an experimental venture for harvesting Management Area scallops, based on our mass of research findings. But the sub-secretary's office stayed silent. We received no reply, and our second request also got no response. By our third try, we got more sophisticated and submitted a joint request from our university and the National Fisheries Service, yet still received no word. What could I say to the fishermen who needed to earn cash? Things were looking desperate, so for the fourth try, I asked our highest Region IV official—the intendente—to file the request on our behalf. What do you think happened?" A wry smile plays across his lips and he shifts position. I shake my head.

Wolfgang continues in a lowered voice as I lean forward, "Again . . . absolutely nothing, all of which was incredible to me, since there was nothing left to research. Puerto Aldea's waters had been mined for data, with more marine biology studies than anywhere else in Chile. And caleta members attended all those workshops. Scallop harvesting, ignoring rules, is a perfectly logical decision by the caleta under these circumstances. But, we won't give up. The system itself needs to evolve."

Wolfgang and his group's sophisticated efforts underscore the impossible position of caletas without sponsorship. A caleta alone can go broke trying to commission the piles of scientific reports and studies, besides lacking lobbying networks. Chilean caleta leaders beginning the Management Areas process soon understood by looking at those caletas being awarded the initial Management Area decrees that each one had biologist allies, usually under the auspices of regional universities, as in Iquique (chapter 7) and Coquimbo Bay (chapter 4). It wasn't a prerequisite; rather it was the de facto practice.

Finally, in an attempt to share power and decision making at the grassroots level, Wolfgang also produced a simplified paper on scallop stock management for caleta fishermen. He, Sergio, and the others realized that despite hours spent attending workshops, the required math and science for Management Area management planning remained beyond the formal training of caleta fishermen. New tools, such as the ones they were proposing, would have to be designed jointly with fishermen down the road.

Nelson and some fishermen and I were sitting on the garden bluff near a man-sized model battleship, a wooden local landmark and local labor reflecting the caleta's Navy obsession. Nelson says, "Though our university friends see that our observations need to be respected and incorporated into biological models, the level of mathematics in management models is far too technical. It's not accessible to our fishermen. But we really do know a lot about our marine life. Policy makers rarely acknowledge our understanding."

A caleta's pattern of local knowledge—the fishermen's tacit nonscientific, in-the-water observation and assessment of meteorological and biological trends—is essential in recreating reliable and acceptable management tools. Wolfgang Stotz has been a major proponent of this at conferences. Sadly, policy makers have been skipping over this wisdom, willing neither to incorporate fishermen's understanding, nor to retool with models the fishermen can comprehend. Policy makers are also discounting the myriad completed biological studies, instead demanding new ones to keep up with the sea's perpetually shifting ecology. Caleta leaders and their biologists cannot win at this unending task.

The Lure of El Niño-Induced Scallop Wealth

During 1995–1996, group optimism degenerated into solo political activism by Nelson as the caleta leader with the university's vital biological assistance from Sergio. Eventually, other caleta members turned on Sergio as the straw dog, blaming him and revoking his caleta membership when they still failed to get their legal Management Area with rights to harvest. This catalyzed the collapse of management efforts just prior to the major El Niño event that induced a wealth of scallops inside the Management Area. Any motivation to obey the government ban on scallop harvesting was ruptured under these sociopolitical and environmental conditions.

Significantly, the caleta's management efforts had collapsed prior to the El Niño-induced scallop riches: scallops multiplying wildly in water with the warmth that also drives the predators to desertion. Official reluctance to grant Management Area status had hindered if not prevented Puerto Aldea from coordinating and strengthening its own management efforts around the Management Area. Gradually, Puerto Aldea people also realized that the entrenched political economic interests in Tongoy Bay's tourism, combined with reluctance on the part of the Navy to allow fishermen marine tenure in Navy waters, had heightened the barriers and weakened a real chance for the Management Area.

In retrospect Nelson tells me, "Caleta members were trapped between choices. They felt they could be either *aventurero o estable*." The translation is like the words sound in English: they could rebel or comply. The caleta saw these two choices, either adventurously start to harvest outside the law, or continue beating their heads into the bureaucratic wall and dissipate nature's bounty. Nelson could sympathize privately with how they felt doubly trapped.

A week later, as I'm helping hold down a billowing net being mended, another caleta member summed up that consensus. He says, "Why should we have continued listening to Nelson and all those Management Area fantasies? We really tried, but the government never truly wants to share managing the sea with the likes of us. How could they expect us to protect and grow those scallops, but not to harvest them? Then the El Niño came like manna from heaven, which only verified our decision to collect. Wouldn't it have been perfectly crazy to let those scallops go to waste?"

A blow strikes the caleta women in particular in 1997, when their best nori buyer, a Korean man, loses his business. Now they are at the mercy of a sole buyer. I learn of this when Susana and some other women involved in the collective nori harvesting that is the heart of women's income are showing me one of the hydroponic gardens around the caleta. Set in the lee of a house, we pass into a separate world enveloped in plastic sheets, safe from the wind's whipping sting of sand and salt. It's an oasis of vivid flower mosaics and tomato plants. We bring some yellow flowers into the house to brighten the table for tea.

A particularly tough and respected woman among us says, "In less than two years, our nori income has just about disappeared. It was the Korean buyer's poor administration. He was buying everything we brought in at good prices, more than he could process. His business collapsed. Nelson, for his part, had relaxed searching for more nori buyers for us because we all thought we had the perfect situation going. Suddenly, we were back to one buyer paying a ridiculously low price."

But by this time, Nelson, like everyone else, is frustrated by the snail-paced Management Area process. By the time El Niño gets underway, Nelson is already spending most of his time living in Coquimbo and working for the wider Coquimbo municipality to develop artisanal fisheries. Technically, he is still head of the Puerto Aldea caleta. Neither Nelson nor the university could possibly have continued inculcating caleta members with Management Area concepts while everyone watches nature producing huge quantities of scallops. The Management Area ideals Nelson had worked to establish up to this point can only fall on deaf ears as long as fishermen are denied the legal right to harvest their Management Area's treasure.

El Niño in Tongoy Bay—Varazón Damages Scallop Beds in Tongoy

In Tongoy Bay, the plethora of scallops was accompanied by destructive storm surges. A not uncommon phenomenon of currents called *varazón* swept through Puerto Aldea's Management Area in September 1998 and ran hard for twelve days, ripping scallops out of their beds and hurling them onto the sand. The Coquimbo newspaper reported that the varazón washed more than 70,000 scallops of all sizes up onto the beach. People were finding scallops tumbled and largely inedible, according to the newspaper. The cause of this particular varazón was El Niño. Two months later, another major El Niño varazón tore through the beds.

Despite being reported in the press as a disaster, it was less so in reality. People from both sides of the bay, Puerto Aldea and Tongoy, descended on the mess and retrieved quantities of scallops that were still edible or even salable, leaving behind those too impregnated with gritty sand. But, in attempting to apply political pressure to finalize the Management Area, Nelson, as the caleta leader, exaggerated the situation to the reporters. Nelson explains his reasoning to me, saying, "Publicity about the wild varazón that left ruined scallops on the beach was true enough, and for a good cause. It really was a complete scene with folks combing the shore for scallops heaped everywhere, tangled in seaweed, some still delicious and others wasted."

These acts of nature underscore how the Management Area divers senselessly lost profit by being blocked from harvesting, and how the caleta's ability to grow institutionally was hampered. However, another facet of the story is local knowledge that a strong El Niño classically stimulates scallop wealth. Tongoy Bay's wild and cultivated populations evidenced this. Scallops are thriving on the heels of the 1997–1998 event. Warm waters cause shellfish, such as scallops, to produce bumper yields.

Indiscriminate Harvesting in Tongoy Rewarded

Ironically, the turn of events in Tongoy Bay followed years of rather indiscriminate harvesting by Puerto Aldea's caleta, which was too demoralized to coordinate and harvest in an organized, restrained fashion while endlessly awaiting Management Area status with rights to harvest. In caleta logic, divers felt quite wise to have preemptively harvested their scallops—albeit illegally—a step ahead of El Niño. They had foreseen high chances for varazón currents with rough seas that accompany a great El Niño. And using local knowledge, they had predicted the event by seeing such classic harbingers as the appearance of warmer water species. From the sheer ecological perspective of biologists, however, it would have been better not to have gathered profit before the varazón, because divers only applied extra pressure on the stock.

Aftermath of the El Niño in Puerto Aldea—Harvesting Nori and the Role of Women

The rough seas of El Niño caused turbidity in Tongoy Bay that limited the growth of nori seaweed in the summer months of 1997–1998. Therefore, Puerto Aldea caleta developed a new fishing schedule in cycles to allow reproduction and grow-out time. Here, members would harvest for eight days and then rest for twenty days. The boats head out around 7:30 AM, each with a diver in wetsuit and the *tele*, who runs both the boat and the roaring hookah compressor, in sweatshirt at the helm. Often for this particular kind of harvest, a teenage helper comes along who earns his 10 percent share of the boat's daily quota.

Traditionally, the daily quota is divided into three parts: boat owner, diver, tele. The diver goes down and stuffs his net bag with up to 80 kilograms of nori, hand-plucking the weed in a way that automatically leaves a bit on the bottom that will regenerate. All the divers are men, save Ana.

Ana, a broad-shouldered young woman, born and raised in Puerto Aldea, wins swimming contests for her secondary school and dives when she's home. I visit Ana's house and find her mother out back weeding in their hothouse garden. She's the woman leading Puerto Aldea's hothouse efforts. She takes me into the main room facing the sea and brings out coffee and tinned cookies.

Pointing to thick paper tacked up on the wall above the table she says, "That's Ana's most recent swimming award." As a pioneer type herself, she's proud of her daughter's unconventional prowess and ability to dive with the best of the men here. It's mid-morning and we look out at boats returning and offloading their nori, which they call chicorea.

Together, we head down the path to the silvery processing shed where other women in bright head scarves and tall rubber boots are carrying in the seaweed from the wharf. These caleta women are working side by side with the buyer's workers. Inside the processing shed we meet five local teenage girls cleaning and ordering the seaweed.

"With a few full days, I earn enough to help my mother and go shop for things like these earrings," says Lydia, showing her style and practicality in helping the family economy.

Floradina: A Teenage Processor

Floradina is another of the teenage processors, and like the others, she will be off to secondary school in Tongoy when the term starts in March. She and her friends wear long white aprons covering their skinny tops and jeans. I'm standing to one side, watching their quick movements organizing the mass of reddish-purple and green sheets of weed onto proper drying racks. Floradina stops for a few minutes' soda break, and I ask her about her father. "He's a diver," she says, "See him over there by the entrance talking to our buyer. The one in blue, that's Chino. Everything we dry, he ships to Japan. He brought us a package of it pressed into paper-thin sheets."

I ask her if she's ever tasted this transformed chicorea as a nori roll around raw fish and rice. She shakes her head, "But my mother boils up cochayuyo in fish soup." I nod, also liking these fist-sized, bulbous tawny heads of a different type of dried seaweed, with long whip tails that unwind once soaked. They're sold in one-headed coiled bundles in local markets. I wondered where her father was born and she said, "He came here from the Totoral area. You know, if you take that track behind the village and go even further from civilization, there's *totoral*—reed grass—perfect for goats. My aunt left too. I stay at her house in Tongoy when I'm back at school."

Population of Puerto Aldea

The local schoolteacher, Emma, is talking to me after hours in an airy classroom decorated with drawings of Navy warships. I ask how many students come here. Emma says, "The school has twenty-six kids total in first through sixth grade." I speculate how many will leave like Floradina for more education. "Last year, half of our four graduates went to Tongoy for middle and high school." We start discussing who will return to raise their own families. Emma pauses, "Half the people here are in their twenties and thirties, bringing up kids. And if half the teenagers stay we'll be good." Emma and the caleta leader differ in their estimates of the resident population: 270 versus 400.

The caleta leader, Nelson, is on the high side as part of his negotiations to win government development. His estimate is part of his political angling for electrification, making the community more amenable to a comfortable lifestyle and building up the caleta's capacity to operate a processing company, aside from attaining official recognition for their Management Area. Another of his politically motivated exaggerations is to present Puerto Aldea as 80 percent native born with historical roots and claims to the area. In reality about half the population moved here when migration was widespread up and down the coast between caletas.

Fishery-Related Jobs and Migration

Looking at current migration, moving away at least temporarily for fisheries-related jobs, I find three couples and four youth from Puerto Aldea working in Tongoy, but nobody is in Coquimbo in fish plants. Some people normally go up to Iquique caletas for the octopus season; some live in other caletas as divers; one is even fishing in the industrial fleet in Iquique. Migrating for non-fisheries work, I find a few away on businesses or working construction part-time and returning periodically to Puerto Aldea.

Nelson suggests an idea for earning income that could hold youth in the community. He suggests, "Have them build and run cabanas for the summer Tongoy tourist rush. But we repeatedly run into the same obstacle—land rights. The Navy always owns the first 80 meters of coast from the high tide mark so we need their permission to build, and they prefer to keep us small over here in their practice area."

Proposed Puerto Aldea Management Area

The official moratorium on wild scallop harvesting in Chile remains in force. Yet Puerto Aldea continues to collect scallops. The proposed Management Area spans ninety hectares and is capable of sustaining nine million scallops. Currently scallops are at one-third that number, but with excellent prospects given the seedlings produced during El Niño to repopulate the area. Of various pre–El Niño projects implemented in efforts to design their Management Area, only the pending scallop project persists because of Nelson's political efforts and university support. Sergio, for example, writes the reports that Nelson sends to the Navy in an attempt to gain approval, and he also produces technical papers for Nelson to use as ammunition on his lobbying trips.

Concept of Natural Resource Use: Wasteful If Scallops Are Allowed to Go Unharvested

In Puerto Aldea's concept of natural resource use, it is wasteful to let scallops grow and perish unharvested, especially when visibly abundant. So Nelson traveled to the capital to request permission for them to harvest 5 percent of the legal-size scallops before another storm could strike. He met with the sub-secretary of fisheries who praised the caleta's work with the university, including their management training. The sub-secretary promised a decision within one week. The week came and went. Even though it was during the scallops' reproductive season, the university wanted the sub-secretary to grant the request. Otherwise, the divers would be forced by economic logic to leave the management track they'd followed for four years.

Pending Management Area

Like Coquimbo Bay, Tongoy Bay has pending Management Areas whose future development and ability to function are in question, susceptible to political and environmental changes that are part of each bay's particular historical experience. The two bays provide us with a good picture of the many factors affecting the likelihood of successfully organizing fisheries in Chile. These include the level of physical isolation of the caleta from other fishermen and markets; the existence of a caleta-university relationship; the level and quality of leadership at the caleta level; and the caleta's informal system of management. Simple variables, such as the typography of the shellfish bed, can greatly alter the ability to conserve and control the stock.

More complicated, intertwined relationships may not be as clear, such as competing hierarchies among the interest groups, each possessing various rights and making different claims. These groups must coordinate among themselves to control resources and thus to participate in the management of their Management Area. We find divers and waders; waders and the informal urban poor, who harvest as a welfare net; divers and their caleta leaders; biologists and caleta leaders; divers and practitioners of other fishery arts like longliners; the Navy and government agencies involved in fisheries; and caletas and tourism interests in both bays. In general, it appears to be easier for a Management Area to unify when the caleta's members are relatively few and interrelated, such as in Puerto Aldea and Peñuelas. Furthermore, the whole process can be stymied by powerful interest groups with other goals, such as the Navy in Tongoy Bay, which is no contest for a little caleta.

Problem of Poaching

The government won't help to control poaching as long as the Management Area's status remains in limbo. Pepe, a caleta diver, has come in from the 8 AM–2 PM shift guarding the Management Area. Peeling off his slicker, hungry and tired, he speaks with general frustration, "It's my boat's day of the month to watch for raiders. We have to go back out tonight from 10 PM to 2 AM when it's much harder to see who's sneaking around. If we don't guard, the pirates prevail."

Pepe tells me what he witnessed on the water today. "We dashed over to a couple of Tongoy boats that were sitting by our border buoys trying to look all calm and innocent. But we traced their divers' air bubbles surfacing at least 100 meters inside our area. It's easy for divers to duck into our area and travel underwater on the air hoses. Who were they fooling? I jumped onboard their boat, shouting at the compressor guy to jerk on the hoses to signal their guys back. When the three divers emerged, each had an empty bag. Must have dumped our scallops on the way up. Ha!"

Although everyone in the bay acknowledges that it is Puerto Aldea's pending Management Area, robbery remains rampant. Pepe describes how a month ago, he and the others specifically set a line of buoys encircling their Management Area so nobody could claim ignorance or innocence. Moreover, they organized a guard system to contain pirate poaching. "We've paired up into thirty-one teams to cover each day of the month. Meeting this schedule is tough, given how life flows, so we have to juggle crew among ourselves to keep all the slots covered."

Pepe explains how urgent this is with the Management Area currently overstocked, offering such a strong temptation. Pepe confides, "Normally, a diver works about five hours to take 500 scallops, but inside our Management Area at the moment, flush with product, you can pick up four times that in half an hour."

So, despite Pepe and the other guards, battles ensue. Caleta Tongoy boats cluster on the Management Area border while Tongoy divers head inside to reap the rich beds. Local press reports Tongoyinos versus Aldeanos, one man bashing another over the head with an oar as Puerto Aldea guys leapt into the boats of the invaders. Puerto Aldea guards have tried confiscating the scallops to use them as evidence in formal complaints to the Tongoy marine government—Gobernación Marítima. This authority issues fines or suspends work permits for five to ten days, so boats caught red-handed surreptitiously slip their haul over the side to skip penalties. If Puerto Aldea's Management Area is ever approved—and hope is flagging all around—the caleta will be able to harvest and market scallops legally, and scallops will no longer accumulate in profusion, madly tempting to the caletas and neighboring pirates alike.

CONCLUSIONS

A charismatic leader has controlled Puerto Aldea caleta, typical of leaders among fishermen. Nelson, founding caleta president, fueled by a sense of social idealism, settled in and took charge with his superior education, buyer status, and understanding of the wider world. During the Pinochet era in particular, in the more isolated Chilean caletas men could carry out their ideals in this way while staying off the political radar.

As leader, Nelson makes the decisions, although three elected directors of the caleta supposedly decide when and how much to harvest within their Management Area. Nelson's caleta vision is to increase stability—to change fishermen from being adventurous—*aventurero.*

Nelson organized one fund for operating costs and another fund for credit, emergency money, and one month's storm subsistence. To keep the scallops safely inside their pending Management Area, and to prevent future losses from another El Niño–derived varazón, Nelson has persuaded members to collect rocks for building anti–storm surge corrals.

Initially, the prospect of the Management Area altered Puerto Aldea's extraction strategies and motivated actions in line with a conservation ethic. Over years of Management Area stalemate in Chile, university biologists often stayed solid allies, as in Puerto Aldea. But many caletas, including Puerto Aldea, were losing trust in ever obtaining legal marine tenure. Puerto Aldea makes a perfect case: because they felt they had proved themselves to no avail, they resumed a level of "subversive" harvesting. How could they afford to act illogically and let abundant mature scallops die in their beds? Why obey strict self-regulation that forbids harvesting just because the Management Area is not formalized through no fault of their own?

EPILOGUE: SHIFTING EXTRACTION STRATEGIES

Biblical in proportion for the communities experiencing changes in their resources, the 1997–1998 El Niño event has brought opposing environmental changes to the nearby bays of Tongoy and Coquimbo. Allowing an extraordinary comparison, ironically El Niño rewarded Puerto Aldea caleta, which had abandoned strict regulation practices, yet devastated the Coquimbo caletas, which had maintained them. El Niño unveiled faces of the powerful players in each situation; it may have been a blessing in disguise for both bays.

Caleta people noticed this opposition between their bays. They watched how macha devastation prompted the government to finalize Management Areas and have enforcement turn an empathetic blind eye toward migration and fishing without permits in Coquimbo Bay. But in Tongoy Bay, El Niño's gift of a scallop cornucopia failed to result in either Management Area status, or in the legal right to sell the plethora of wild scallops. Nelson keeps a ray of hope for the legalization of their Management Area. He believes the Management Area could bring everyone back into the fold of conservation and legality, allowing caleta members to manage and market more cohesively and efficiently.

An admiral arrived for an impromptu visit during the El Niño to discuss the future marine tenure rights of Puerto Aldea. Though the caleta complies with war games and is willing to clear the Management Area of their marker buoys, the Navy is reluctant to relinquish total rule of the seas. Behind the scenes, they are recalcitrant about permitting the Management Area in their half of Tongoy Bay, war-game waters. Stymied by the Navy's fear, nature has been the generous one toward the Management Area this time.

Two key plans for the fate of Puerto Aldea awaited the signature of the Navy: one is the Management Area decree; the other was permission to run electricity over a half-kilometer stretch of land to the village. Within the community in late 1998, a new ray of enthusiasm stirs around the arrival of electricity that people believe will fulfill the potential of turning the shiny processing shed into the caleta's small enterprise. They call the shed *galpón*, their "horse," to ride to economic abundance.

As Nelson is showing me around the shed, he confides that electricity is not the panacea. His community doesn't have trained people to run such an enterprise, from the engineer for the freezer technology and plant to health requirement experts to accountants and marketers. To formalize a full-blown business they will have to import people with these skills and have their youth learn them over time.

Meanwhile, on the eve of electricity, the shed by the wharf, completed two years ago, is locked. Nori is hanging to dry in the conveniently large sheltered space, yet the infrastructure to establish a processing plant for caleta harvests sits idle: icebox, freezer, and vacuum packer. Locals speak of their hopes for the shed becoming a place where teenage and married women can earn money, ideally only working half days to have time for their families. The other half of the economic plan foresees everyone receiving higher prices for their products. Their product ideas so far include vacuum frozen bags of *piure* and scallops destined for the national market. Under Nelson's influence as a professional buyer, the caleta desires to bypass selling even scallops to the aquaculture plants in Tongoy, and become buyers themselves, by going to Tongoy and other neighboring caletas.

LETTER: THE MYSTERIOUS HERMIT OF SALAR DE HUASCO

Dear Lois,

Wishing Ms. Wily well and purring in your magical Tyring Ham Valley. Last weekend at Puerto Aldea in celebration for the arrival of electricity, I'm helping the women lay out shellfish delicacies for urban dignitaries due any minute. They pull up in several cars outside the one-room community center transformed into fiesta mode. Big plates of food stretch out on tables and crepe paper streamers fly. Folks crowd in, children dashing around. Once everyone settles into folding chairs, Nelson, Puerto Aldea's leader, lays in with the welcome speech, climaxing to applause for the old adage, "You give a man a fish for one day, but you can teach a man to fish for a lifetime." After more political hoopla about progress, Nelson and the assembly seeking popular votes throw the switch. The new electric streetlights up on poles posted around the village compete with stars starting to come out.

Now, I've returned to northern Chile, land of dune dragons. High above Iquique's fishing villages, in the cactus altiplano, I'd heard of Conrado, a "disciple of Einstein," a hermit astronomer-mathematician, and want to meet him. Dale, my Feldenkrais body-awareness friend, has just arrived by overnight bus from Peru, so I rent a pickup for our expedition to converse with the hermit. We head for the settlement closest to Conrado's haunts, climbing through high moonscape in blue air, ears popping, finally entering the lemon-laden oasis of Pica.

At the general store across from Pica's hot spring, I luck into finding Toño, the Aymara owner who says he knew Conrado, and the Aymara living in the pampa who'd looked out for him, bringing him bread and fruit for thirty years. But we just missed him! Conrado died three years ago. Toño draws me a map and explains the dirt track assent that will take us far higher the next morning to Conrado's place, Salar de Huasco.

Here we feel light-headed and leaded down by altitude, walking around the remains of Conrado's settlement. He'd dug down four feet for earth's warmth and rearranged stones from ancient camelid corrals to build walls shielding him from bitter winds in his home of wild guanacos. He kept no dog as pet, but a mountain vizcacha—a rabbitlike creature with a long tail living in rock clefts in colonies—would come sit by him and eat sugar crystal offerings. Knowing the heavens, discussing how time is a circle and how negative time allowed humans to reach the moon, Conrado placed rocks in a circle, matching clock time within minutes, amazing his Aymara visitors. Some say he talked of angels, others say he wrote diaries about the universe full of equations. Upon his death, a Santiago lawyer friend took all his papers, but when he died suddenly, his law partner son tossed out everything extraneous in their office.

What did Conrado think about? He called talking about his family "taboo." Supposedly, his French wife left him, and a son flew in the clouds as a pilot. When people bothered him with "absurd" questions, he would pick up and move to an even more remote semi-troglodyte camp, one he'd built within a well. In Conrado's can middens, I find pressed and oxidized Cargo Art spirits now keeping me company. They'll come home to Miami's sea-level altitude.

We watch pink flamingoes feeding in the distance on the *salar*—salt flat—and jump and climb about the rocks and rock walls. Leaning into the oxygen-thin stiff wind, I pat a *yareta's* ancient rough hard emerald head.

Yareta resembles a large brain coral head and grows equally slowly in terms of millimeters per year. It's one of the treasures seen only at this rare altitude. Though it's forbidden to harvest unless already dead, in southern Bolivia I'd met poor campesinos living as virtual hermits, set up by brokers to harvest the hard woody plant. Occasionally, the brokers truck in food in exchange for the cut yareta. It is burned in ovens to create charcoal for sale to Brazil. What a sad end, similar to slow-growing coral being broken off and burned to become lime for building material or as the catalyst for chewing betel nut.

Back down in the oasis among Conrado's Aymara friends, I learn another folk theory now being passed around: the earth's tilting causes El Niño. Aha! Another friend adds that Chile's deserts will grow green in the new millennium, as the earth tilts some more. Flowering deserts are to become normal in their future . . .

Warm hugs and happy morel hunting in the woods with Murray,

Sarah

Chapter Six

The Goliath in Iquique

Industrial Fisheries

> "Say *obreros*, not *trabajadores*!" —Iquique union leaders, who keep urging me to switch my word choice for workers

Language is critical in declaring oneself pro-proletariat, *obrero* signifying solidarity of the Left. I understand the weight of these words, given the bloody labor history in Iquique. I need to dance carefully, depending upon whether I'm speaking to labor leaders or managers.

LIFE BY CERRO DRAGÓN

Iquique is quintessential Region I, the northern reaches of Chile, the desert coast running south of Peru (figure 6.1). This is my annual visit, the town melding into sands by the sea is even more dreamlike than I'd remembered. Many buildings are painted brilliant tones in sharp contrast to the cold, dark sea on one side and the town's mountain margins of bizarre, rocky landscape colored by mineral shades, clays, and guano. Lining both sides of Baquedano, the main street leading from the ocean through to the central plaza, are recently renovated, three- and four-story wooden buildings with lacy trim, full frontal balconies on each floor sporting flagpoles with flags snapping in the breeze. Although we're under a blazing summer sun, winds either sweep down from the high desert or gust in off the sea cooling the city.

The plaza pivots around the Peruvian-era clock tower, built before Peru lost this region to Chile in 1883. It's occupied by ravenesque black birds that soar above people lounging on benches in the shade of towering palms that could easily date from the same era. Children nearby are shouting and run-

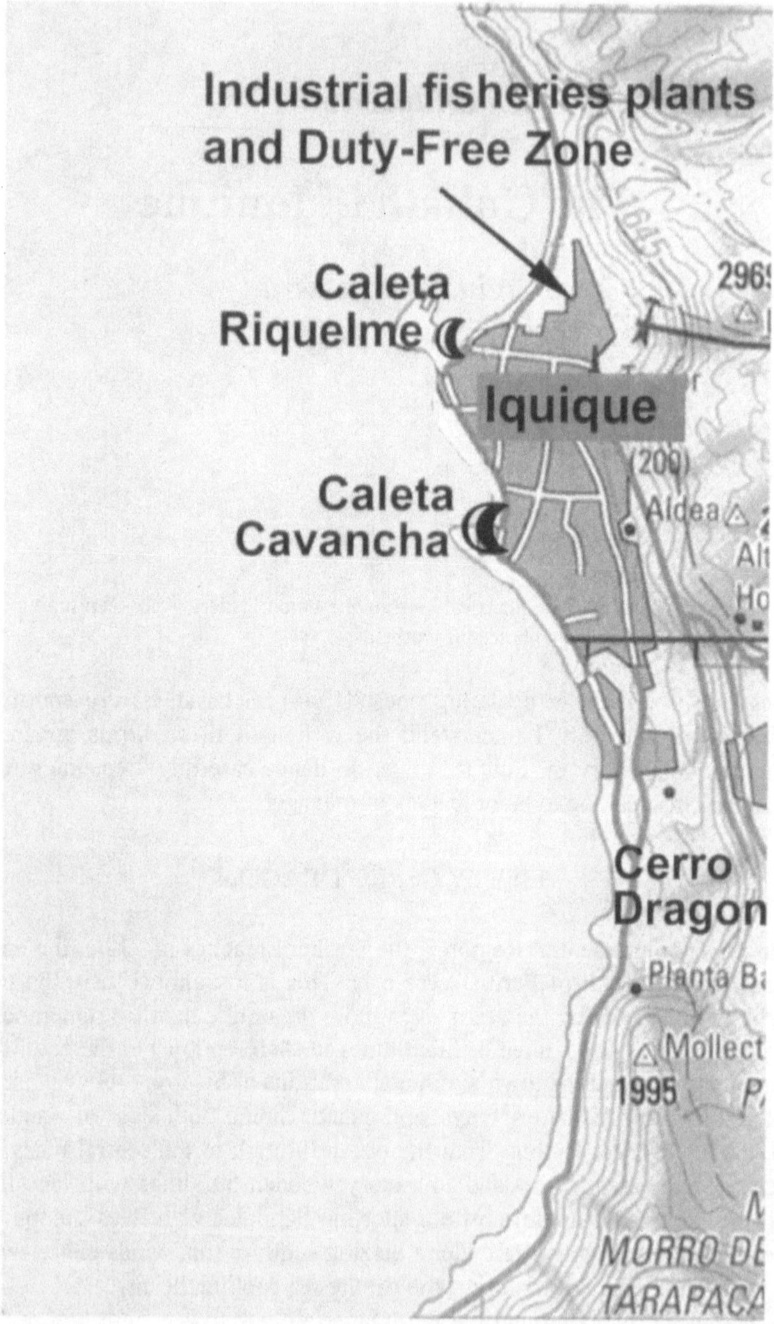

Figure 6.1. Urban Iquique map. Adapted from base map from U.S. National Imagery and Mapping Agency, Perry-Castañeda Library.

ning up and down the wide slope of steps in front of the theater at this end of the plaza. The theater and the other historic buildings are all airy, high-ceilinged affairs of wood. According to common oral legend, men in this treeless region constructed these buildings by dismantling ships for their Oregon pine.

Locals are openhearted toward the few visitors. Gringos—Euro-American types—are still a novelty in the 1990s, despite Iquique's growing reputation as a beach town and a surfer's haven since the 1982–1983 El Niño *del siglo*—of the century—killed off the nearshore giant kelp beds, opening up conditions for great waves.

Above our modest rented house, on a street of matching two-story cinder-block homes, looms the symbol of the city, Cerro Dragón—Dragon Hill—an enormous golden dune. She subtly shifts sand postures in the continuous winds. Late afternoon light glows along her enormous tail and up the ridge of her spine, as she guards Iquique's south side. Below our working-class neighborhood, the streets drop to the sea. I often slide down the final precipitous sandy path to reach the surf line. Here, I do tai chi on the wide hard expanses of damp sand. I dodge the surge with sandpipers racing through the foam, gulls rafting on midnight blue waves.

High in the altiplano, at altitudes far above the abandoned nitrate mines that once fueled this port economy, two new world-class copper mines operate (figure 6.2). But industrial and artisanal fisheries are Iquique's visible economy along with nascent tourism.

ETHNOGRAPHIC STORY AND EMERGENCE OF GLOBAL THEMES

Rise of Iquique's Industrial Fisheries

Iquique's fishing industry was jump-started in 1960, when Chile's President Jorge Alessandri (1958–1964) enacted the law to create industrial free-trade zones at either end of the country. He was reacting to the death of the nitrate-mining industry in northern Chile caused by the rising market dominance of German synthetic nitrates in the late 1950s. This industrial undoing left thousands of miners jobless. Many poured down to the coast in search of new work, heightening social unrest. In 1959, Iquique residents held the famous town-wide protest remembered as Banderas Negras—named after the black flags that they draped from the ubiquitous balconies to demonstrate despair and outrage. Besides the lack of work, they were angry about the pitted dirt roads, scarce electricity, and an inadequate supply of water, so critical in this desert climate. In other words, they were angry about the lack of infrastructure required for a decent livelihood.

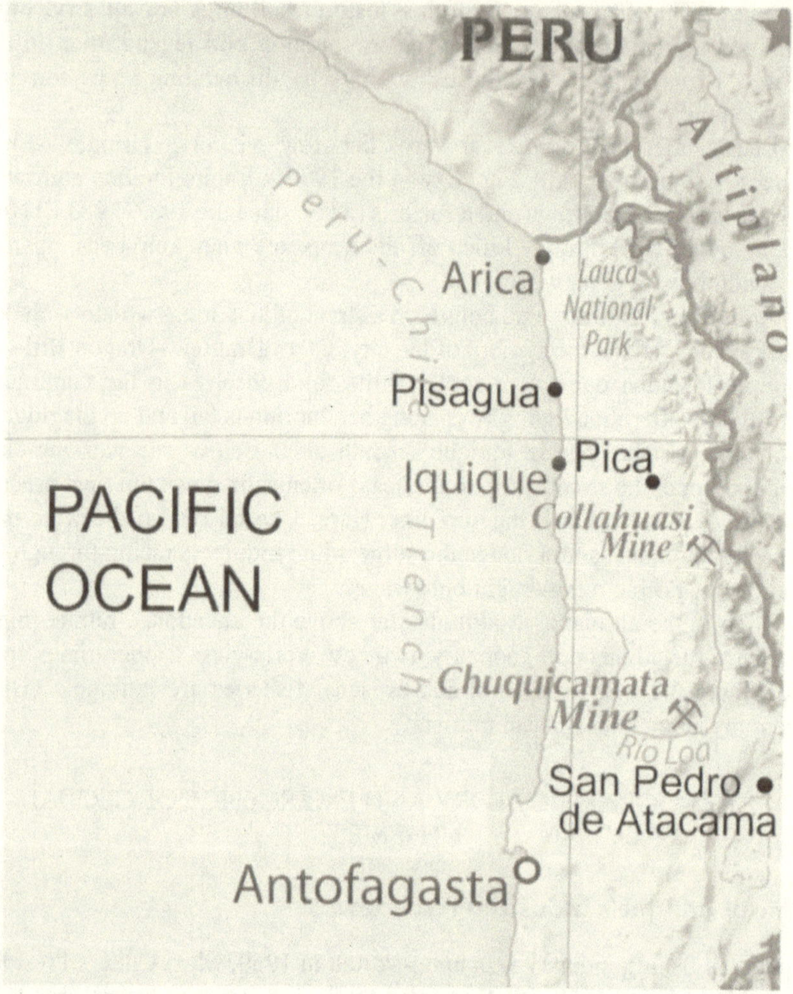

Figure 6.2. Northern Chile and the mines map. Adapted from base map from https://www.cia.gov/library/publications/the-world-factbook/.

Specifically to encourage start-ups in the fishing industry, the government organized and revamped their credit system. On paper, many fishing companies applied and received credit, using it to commence preliminary operations. Only half of the companies given credit actually created production jobs. Not surprisingly, the majority of loans await collection forty years later.

Investors began constructing fish factories in Iquique during the early 1960s. They saw Iquique as advantageously situated for the capture of the fishmeal and fish-oil species, sardines and anchovies. Iquique vessels could fan out to fish, while vessels in Arica, abutting Peru, legally could only head southward. Iquique's first two factories had been canneries in Caleta Cavancha, then a fishing village with fishermen's houses out on the peninsula surrounded by water on three sides. Bonito and tuna were the favored fish to can. Soon, a third cannery started operating right in Iquique. Canning was followed by the production of fishmeal—*harina de pescado*—which endures as the strong arm of Iquique's processing industry.

Birth of the Industrial Fleet

Rico, a native of Arica was drawn to Iquique during that early period of industrial production. He's sitting opposite me at the Chinese restaurant where we're having lunch, thumbing through his collection of old postcards, holding out pictures to me across the soy sauce and chop suey. He leans over to point out details in these early images of plant chimneys, tall plumes rising smoky above a town held between cliff face and sea.

We're becoming friends through a mutual passion—photographing caleta life. Rico's wife, Sofi, can make tender seafood wonton as good as what we're being served. She's Chinese Chilean, representative of so many people whose ancestors were brought into Region I as contract mine laborers to do the harsh work of extracting nitrate and salt. Over several generations, the Chinese rose into business and intermarried, diversifying the town's cultural mix and lending this region's food its distinctive Chinese Chilean flavor.

Rico started out earning industrial wages, working his way into the artisanal fleet, saving for his own boat while crewing to gain enough knowledge to become a captain like his uncle back in Arica. Like me, he captures scenes in photos that he realizes will look as antiquated as his postcards in another half century. Rico says, "Chile's government began touting the fishing industry as the *salvavidas*—life saver—of Iquique. People like me with no wages were drawn by *el olor del dólar*—the smell of the dollar—permeating streets, houses, and even our clothes." Rico metaphorically likens the allure of income to the stench emanating from the numerous fishmeal plants; it's a classic Iquique pun.

I ask about his fellow plant workers back in those days. Rico explains, "Ex-miners coming down from the pampas were the largest group. But word flies during tough times and soon jobless guys like me were migrating here from all over the country. We made rough little places to live up on those slopes above Iquique's wharfs, shipyards, and plants." He shows me another postcard of workers in baggy pants and high rubber boots lined up outside a plant between shifts. Everyone, from crew to shore labor, worked hard to produce the steady supply required by the export market. The high availability of both labor and pelagic baitfish favored Iquique's vertically organized industry, where the fleet and the factory were parts of the same company.

Powerful purse seine nets—averaging 900 meters wide and 120 meters deep—and sometimes as deep as 400 meters—swept the seas, seining in the fisheries art called *cerco*. Along with the dramatic water temperature rise during strong El Niño events, purse seining was affecting stocks, but this was ignored by all.

Positive Start on Industrial and Artisanal Fishing Relations

Artisanal fishermen were not left out of the initial rush to industrialize fisheries. Caleta Cavancha's fishermen upped their hunt for good canning fish, such as tuna and bonito, for the two new canneries—*conserveras*—in their midst (figure 6.3). Canneries owned the vessels, lending them to fishermen in exchange for a percentage of the catch. The canneries recruited women to process the fish, including wives of fishermen. These workers earned the nickname *chamaya*—smelly little fish—from going home impregnated with fish stench.

In fish-processing plants that I've observed from American Samoa and Spain to Ecuador, Peru, and Chile, there tends to be a strong gender preference for hiring women as processors. Women are perceived by plant manag-

Figure 6.3. Caleta Cavancha boat offloading catch. Courtesy of the author.

ers as being better adapted to meticulous manual seafood cleaning and processing, besides being more willing to stick out jobs for the sake of caring for their children. In keeping with technological changes modernizing fishmeal plants, the shift to machine labor replaced women with men on the processing floor; now only men worked the floor, and a few women held secretarial, administrative positions.

Consequences of El Niño of 1964–1965: Seesaw from Anchovy to Sardine, and Day to Night

Fishing in Larger Vessels

The first blow to Iquique's new industry came from the big El Niño event of 1964–1965, when anchovy stocks disappeared. By 1965, the stock crash resulted in a number of companies closing and subsequent unemployment. This was a foretaste of future chaos from disappearing anchovy and/or sardines. Once again, workers in Iquique flew black flags in protest. Empty vessels that had targeted anchovy lined Iquique's docks, Iquique's first generation of industrial purse seiners. At this point, their vessels—which were wider and shorter, with the bridge house in the bow—had become obsolete. These vessels, sometimes accompanied by the mobile skilled crew, were auctioned off globally, ending up in Mexico, Panama, and even South Africa.

A profusion of Iquique companies dropped out. But, six top companies emerged strong and able to persist during these tough three years as stocks recuperated. They were Angelini, Tarapaca, Coloso, Guanaye, Indo, and Iquique. Their successful survival strategy was a dual combination of (1) shifting technology to night fishing for *sardina española*—sardines—and (2) moving their fleets farther offshore to where this new target species was congregating. To put this strategy into operation, the industry needed new vessels to hunt sardines further offshore.

By 1969, the post–El Niño fishing industry in Iquique had suffered a 50 percent overall reduction. At the 1964 peak, Iquique supported twenty-eight fishing-related industries, including 1,800 crew purse seining, and twenty-two shoreside plants employing 8,000 workers.

Post–El Niño, Iquique only had twelve plants and 4,000 shore workers; crew was also down 50 percent. Soon the unemployment and investment woes were exacerbated by the general Latin American economic crisis of the early 1970s, as well as Chile's politically induced economic turmoil.

Crash of the Anchovy and Allende: Swing from Socialism to Pinochet Capitalism and El Niño 1972–1973

In 1970 Salvador Allende, Chile's newly elected socialist president, responded to grave unemployment by nationalizing the industrial fishery. To create jobs, plants were forced to employ 80 percent more workers than necessary. Work was duplicated and, as with scandals in the fishing co-ops, workers often were never properly trained. This inefficient labor force was a source of great controversy.

But the real fisheries crisis was the second collapse of the newly reemerged anchovy stock. Again, the cause was a big El Niño event during the 1972–1973 summer season: the anchovy catch plummeted 80 percent in one year. But once again, the industry was rescued by a simultaneous seesawing in fishmeal species. Anchovies fell and sardines stocks offshore soared.

"The fall of Allende and the fall of anchovy," goes a local rhyme recording events of 1973 when both crashed. Pinochet's military ousted Allende. Naomi Klein (2007) sharply outlines the details of the coup, with the Chicago Boys' adventures in extreme capitalism reengineering Chilean society. Combined with the revival of the Latin American macroeconomy, neoliberals brought on the reversal of prior policies. Fisheries cooperatives were greatly diminished with the shift to massive reprivatization of the industry and the neoliberal emphasis on high production for export revenue. Cooperative funds and shore equipment were confiscated. Pinochet's military force enabled much of this change, including the cruel tactics that silenced labor leaders.

For example, Iquique's fisheries co-op president, Fermin Tabilo, was arrested under Pinochet, like numerous labor leaders in both artisanal and industrial sectors in Iquique. Tabilo was shipped to the infamous coastal prison town of Pisagua. Luckily, after eight months, he was released. But he never spoke of his ordeal or returned to organizing fishermen.

Labor Historian's Story

At the beginning of General Pinochet's coup in 1973, Iquique became a hub of military roundups. Pinochet had been posted in Iquique's major military base, and his military spies had already drawn up lists of subversives. Operations swept down on Iquique's labor leaders and deposited them in Pisagua's natural prison in the coastal desert (Meltzoff et al. 2001). The Escuela de Santa Maria Massacre on 21 December 1907 had turned Iquique into a highly contested and symbolic place in Chilean labor history (figure 6.4). Chilean labor considers it the turning point against their homegrown version of the global labor movement. In pro-labor offices around town, I can usually spot one of the posters from the famous old movie dramatizing Santa Maria.

An estimated 26,000 workers walked down from the nitrate mines up in the pampa to protest peacefully for a few pennies more in wages. Workers and their families poured through the grand main street of Baquedano as onlookers looked down at them from the balconies. Gathered at the Santa Maria School, a general sent by the mine owners told them to go back to the mines empty-handed, but in a non sequitur, the military opened fire, killing at least 3,600.

Heading to a meeting at the central library with Guillermo Ross-Murray, a local labor historian, Cantata de la Santa Maria is playing on a taxi's tape deck. The driver tells me he is currently laid off from crewing on an industrial vessel until the fish stocks revive. The music expresses his strong labor sentiments. I locate Guillermo Ross-Murray—who collaborated with Lessie

Figure 6.4. Pilgrimage to Pinochet's victims' mass grave in Pisagua, where relatives recite Pablo Neruda and wave banners for President Allende and Victor Jara, the Left's political folk singer. Courtesy of the author.

Jo-Frazier (2007) on local labor history—in his haunt high up in the newspaper archives. He's a small, rumpled, sweet man in thick glasses. As we walk out for lunch, Ross-Murray and a number of secondary students milling around the entrance hail each other; he's their guide to the library stacks who makes them come alive. Ross-Murray leads me some blocks away from the library behind the central plaza at the end of Calle Baquedano. He wants to show me the site of Santa Maria. The school is housed in an ordinary blue building with a simple commemorative plaque.

Then my new friend takes me into a restaurant bustling with workers bordering one of the big open-air food markets. On the walls among lots of historical photos I see the movie poster. Ross-Murray orders us Iquique's fresh fish soup of the day. At parting, he presents me with my own tape of Cantata de la Santa Maria.

Under Pinochet New Credit Is Available to Develop Industrial Fishing Infrastructure

Fisheries once again began to expand. In tandem with Pinochet's privatized, export-driven economic policies, a freezer plant was introduced in Iquique, enabling burgeoning export from 1980 onward. Fisheries markets globalized, thanks not only to technological advances in industrial cold storage and transportation, but also to Chile's neoliberal export-focused economic policies.

Iquique tripled in population from the early 1970s to the late 1990s. A general shift in diet preference toward fish also spurred industrial and artisanal fishing. The Chilean government consulted in varying degrees with university marine scientists in developing top-down regulatory measures to address what they saw as out-of-control levels of fishing. In 1986, they imposed limited entry—*acceso limitado*—for anchovies, horse mackerel, and sardines in the industrial fleets.

The majority of industrial fisheries managers (like other business-oriented Chileans, who tend to be pro-Pinochet) have attributed Chile's economic advantages over other Latin American countries to the way they pursue business. They point out their relative bureaucratic honesty, along the lines of English ideals, and they note how government makes business credit available.

Paolo, the Fleet Manager

On the topic of credit for keeping abreast of fisheries change, Paolo, a long-term fleet manager for one of the big six companies, invites me to visit him at his office. It's a long interior room hung with photos of him and his race car, as well as classic and grand new vessels bought through government perks and assistance to the industry. He, himself, has invested in buying up and running a couple of vessels retired from the industrial fleet. Originally, he'd been a noted captain with that intuitively refined skipper sense, referred to as the "skipper effect" (Pálsson, 1994) of where to hunt. Paolo had been the kind of captain that others would follow to discover his fishing grounds.

With a wave of his hand to redirect the conversation to the industry, he offers the example of continuing government credit. He says, "Credit lines here resumed quickly, with Pinochet targeting development of our industrial fishing infrastructure." We're at his desk now, flipping through stacks of his company's bound catch statistics and vessel lists. He then lays out current weather maps predicting surface water temperatures. Paolo describes how these temperature data combine with vessel speed and the company's spotter-plane reports to give him more information with which to decide where to direct his fleet captains. Fleet managers must keep alert to environmental changes to be able to survive the inevitable flux of stocks. Playing off his own intuition, these data merge to provide him with an edge in the race to reaching the bait schools first.

Thanks to taking out his own credit, Paolo works on the side, commanding his own vessels that are part of the Suzuki (semi-industrial) fleet. This fleet is rooted in the industrial sector's shift to swifter vessels that have doubled in size. With Pinochet and sardines, they sold off some of the older, smaller, slower vessels to owners such as Paolo in Iquique and in Coquimbo. The ability of such individuals to also receive government credit to invest at this new semi-industrial level allowed a number of highly skilled captains from both the industrial and artisanal fleets to become *armadores*—vessel owners who direct operations from land. Outside investors bought into this fleet too, angling for the lucrative fishmeal production. Together, they formed the semi-industrial Suzuki fleet.

Earlier, Paolo had authorized the front gatehouse guards to fit me out with a blue hard hat in exchange for my passport. Then I can pass into this industrial fisheries arena, walking through a labyrinth of yards, skirting the various plants and sheds, until I've reached the enormous net shed where his office lies. Upon escorting me out, Paolo allows me to explore his dominion, the mounds of worn purse seines heaped and stretched across the vast shed floor. Walking on top of the net carpeting in light rays shooting down from high windows, I watch and photograph contract workers also in hard hats. They're in clusters hunched on low stools, cutting out torn mesh and knotting in new nylon rope or whole mesh sections.

Once outside, we dodge forklifts ferrying plant equipment around the yard. Parting at the gatehouse, I see an old red setter dog lying bloody. "Vanesa!" cries Paolo, striding over to her. "Vanesa lives here, but lately she's been getting hit because her hearing's gone." She stopped being able to navigate the traffic in her environment. Paolo says he'd been anticipating this end.

As a mandatory route for accessing government credit and benefits, many caletas reorganized cooperatives into syndicates, and joined national artisanal fisheries associations. This is a common theme pointed out by this book, where we see fishermen correspondingly organizing to take advantage of policy offers. The caletas, like the industry, needed to invest to target new species and keep pace with changing markets. For example, watch Caleta Cavancha. Cavancha fishermen had been selling 80 percent of their catch to their co-op. When co-ops were gutted by Pinochet, they turned to sell directly to the industrial markets, especially to the canneries. But around 1980, Chilean cannery production fell as more competitive Peruvian and Asian canneries began turning out less expensive products.

Fishmeal was gaining the stage as Chile's main fisheries industry. Fishmeal plants were less risky than canneries because of lower labor requirements and less machinery. Cavancha fishermen were responding to market changes—another common theme in fisheries where the market creates and motivates the catch. Within three years, the caleta had formed a syndicate to qualify for government credit, and also invested the credit in new fishing technology allowing them to target the fishmeal species needed by the thriving plants.

Iquique becomes a Duty-Free Trade Zone

The Duty-Free Trade Zone—*Zona Franca*—in Iquique has been another major economic booster, in addition to credit. Here, the Chilean government has offered relatively constant planning and assistance that has helped Iquique's business, including the fisheries. In government efforts to boost the economy at the two extremes of this elongated nation, they created a Duty-Free Trade Zone at each end of the country. The Duty-Free Trade Zone came into full swing under Pinochet, who favored his old military haunt of Iquique to become the northern zone.

By the early 1980s, Iquique had become the counterpoint to Punta Arenas, the port of gales in the far south. Iquique replaced Arica, which is precariously placed on the Peruvian border, in case of military conflict. Iquique's rise to Duty-Free Trade Zone translated into big money for business development. For instance, Iquique quickly became the designated place to buy machinery and used vehicles shipped in from North America.

As a perk for living in the extreme Region I, residents are allowed to buy the used vehicles tax free for their own use, not for resale elsewhere. It's a boon that fishermen and workers in other regions view as a wonderful luxury among their Iquique counterparts.

Industrial Boom-to-Bust Cycles through the Eyes of a Net Contratista

My friend Maritza is a marine biologist at the local university. She consults with caletas on making their Management Areas and developing aquaculture. We've been meeting frequently. She often brings me to her house for home-cooked seafood. The conversation flows to how nets are at the heart of Iquique fisheries. The entire industrial fishing sector and the urban artisanal fisheries revolve around the use, design, repair, and trade of nets. To learn firsthand, Maritza wants to introduce me to her husband, Pepino, who owns one of the three net specialist companies in Iquique. So, one weekend, she invites me to the Pirate Club, their private beach club south of town.

¡Todopara los amigos, la ley para los enemigos!—Anything for your friends, the law for your enemy—goes the widespread Spanish proverb; and the Pirates have been able to buy beachfront through their local Navy friends. Recall that in Chile the Navy controls the coastal public domain: the shorefront up to eighty feet above the high tide mark. These influential Iquique Pirates, interrelated through fisheries businesses and management, and the Navy are building a set of weekend fiesta houses up on stilts in the sand. They seem to share a love of maritime memorabilia, so their wooden decks are festooned with everything from old anchors to signal flags flapping in the sea breeze.

On Pepino's invitation, the next week I locate his office on the ground floor of one of Iquique's historic houses of Oregon pine. Old ship riggings are draped like mobiles from the high ceilings. A wooden dory with oars floats in the middle of the floor space. And on his desk an antique brass diving helmet keeps Pepino company like a talking head.

Pepino says, "My introduction to fishing began in 1980, when, like so many other Chileans, I was lured to Iquique by stories of riches being made during the height of the industrial anchovy boom in the late 1970s."

I'm visiting him to learn about how his work evolved in the industry. Pepino describes to me how he worked his way to captaining an Iquique vessel for one year, before returning to shore as harbor chief for the company Guanaye's industrial fleet. After a decade as a harbor chief, Pepino altered his industrial role by joining a net company, Nausicaa. Pepino's title today is fisheries engineer—*ingeniero pesquero*, professionally signifying respect and/or schooling.

Pepino says, "Nausicaa is aptly named after a Greek myth in which the daughter of King Alcinous rescues the stranded Odysseus from a wrathful Poseidon and helps him set sail again." Nausicaa was formed to service nets for the growing industrial fleet that had learned the hard way the direct economic relationship between fish dynamics and net dynamics. The onset of the erratic anchovy and sardine flip-flops in the 1980s, attributed to El Niño events, had driven this lesson home.

Nets as Heart of the Industry

We start discussing the essential business of the nets, from the needs for rapid redesign and planning wrought by El Niño events, to the constant repair work and the industry's general economic shift to contract labor. With this transition, independent companies employ the workers, leaving the fishing industry free of these labor obligations.

"It's important to understand the difference between the two main types of industrial nets," says Pepino, holding up mesh samples for me to examine. "Each type is specialized, one for anchovies, the other for sardines, our two main fishmeal species."

I see that anchovy nets are woven with smaller holes of about half-inch mesh—*malla*. Sardine nets use a mesh size that's about three times larger because sardines are both larger and faster. "Sardine nets cannot capture anchovies. They swim easily right through the mesh. Anchovy nets with the smaller mesh can contain sardines; however, the rapid sardines require a lighter, less dense net that can be dragged to keep pace with them," says Pepino.

I ask Pepino about the transition to relying on net companies as the inherent erratic nature of the fish stocks became evident. "Yes, I jumped into the heart of fisheries, specializing in nets," says Pepino. "The fishing companies found it more economical to switch to us net contractors—*contratistas de redes*—part of the basic trend toward contract labor." Presumably, Pepino and others can operate more competitively and efficiently.

On Nets and Sea Lions

Pepino drives me in his large, sleek jeep out to see his operation on the ground. We pull inside the walls of the wide net yard high on the southern edge of town abutting bluffs. With a huge geoglyph of a whale's eye high above, weavers patch together mesh patterns under the watch of their net leader. The whale is a symbolic reminder from Greenpeace of the former whale factory on the coast below us. They created the geoglyph with the pre-Columbian technique of placing rocks to draw huge figures to be viewed from afar. This modern whale commemorates all of its predecessors processed at that former factory. Now, everyone traveling the road south to the airport and the rural caletas, is watched by the whale.

Pepino leans over one of the frayed sardine nets and says, "Sardines with their extra speed and feistiness beat up nets far faster than anchovies."

I ask about small-scale artisanal nets that the fishermen have to repair. "Sea lions are their problem. Some shoot at sea lions, which rip into nets to chew up fish." When I ask about laws protecting sea lions, Pepino nods, "There's a moratorium on killing them, yet it's about to be lifted. My wife's university thinks there're about 45,000 sea lions in our region." He pauses, "But artisanal fishermen have never stopped hunting them, because sea lion oil and meat are aphrodisiacs, and now Asians secretly pay a high price for *cojones*—testicles—for their medicinal market. I figure there'll be a lot more sea lions killed once the hunt is back out in the open."

I'm thinking of how the classic coastal competition between sea lions and fishermen is about to be escalated by an illicit export market. "Markets create the catch," is the truth in fisheries. And my thoughts stray to Pisagua and the enormous sea-lionesque man by their wharf selling his bottles of sea lion oil for virility.

Net Leaders

Pepino's net weavers are guided in technique and procedure by their professional leader—the *capataz*. Before, the industry fishing companies each kept a chief of nets—*jefe de redes*—who made the final net-design decisions. Then, alterations in net design could be suggested by net leaders as well as by the captain of a fishing vessel. Today's net leader is responsible for knowing both how to repair the massive nets by sewing new sections into worn-out or ripped sections, and how to redesign their size and shape, which is generally done by adding zigzags in appropriate places to deepen the net.

Most net leaders have learned their trade from experience, not schooling. But lately, some of the net leaders are fisheries engineers with university specialization in net design. They have broken in on the scene, using computers for three-dimensional design, working with the concept that various bodies—*cuerpos*—of a net can be quickly cut, added to, or subtracted from, and refitted to create the most efficient shape. So, Pepino's company must offer such up-to-date expertise. Pepino sighs and says, "A growing number of captains, like the net experts, are school trained. Those with formal education act far less humble, and are much more likely to order us on how to reconfigure their nets."

Remembering the 1982–1983 El Niño and Its Aftermath

Before the giant 1982–1983 El Niño struck, the anchovy abundance once again was enriching the Iquique region. Pepino gazes up from the nets into the geoglyph eye of the whale and says, "There were mountains of anchovies, and absolutely everyone went crazy over anchovies—*todo el mundo estaba loco con las anchovetas*. At the start of this El Niño, the anchovies ran thick along the shore, but as the warm waters reached the coast, the concentrations disappeared."

In the classic seesaw pattern, shortly after the event, sardines appeared and swam in great schools, replacing the anchovy abundance. El Niño events have been kind to industrial fisheries in northern Chile.

Compared to other El Niño events, the 1982–1983 El Niño phenomenon has had the most drastic impact on fisheries. For the artisanal sector, the algae ecosystems and many species of inshore shellfish around Iquique perished. Meanwhile, more tropical species, such as dolphinfish, swam down from more equatorial waters, following the warmer water. Northern Chilean artisanal fishermen began catching them and interpreting their appearance as the in-the-water sign signaling a big event. Artisanal fishing renewed.

But sardines, the industrial target species, were displaying astonishing behavior. Pepino and I sit down in the shade of a net shed. His eyes beaming, he draws squares in the sand with his finger. "Iquique radio described the sardine schooling pattern as moving in solid blocks of fish— *bloques de pescado*. Fishing peaked at night, when enormous schools flew like comets through the sea, each with a brilliant head followed by a tail of gleaming nighttime phosphorescence. Phenomenal!

"If the whole comet shot into the net, that vessel would capsize and sink. One company back then lost three vessels, the captains surrounding too many fish at once. My old company, Guanaye, lost two vessels. They actually flipped from the excessive weight and power of those sardines. The main problem was how to split and partially encircle these extraordinary 500- to 1,000-ton schools of high-action fish." Over the next five years after El Niño, this wonderful sardine abundance continued to enrich everyone. Catch efficiency only increased as all the vessels traded in anchovy nets for the more maneuverable, larger-meshed sardine nets.

Industry Caught Off Guard in 1987 El Niño's Seesaw as Sardines Drop and Anchovies Rise

To the industry's dismay, sardine stocks plummeted and anchovies soared again in 1987. However, it was too late to amend the grave error of having switched completely over to sardine nets. The industry had also shifted up into more powerful vessels, double the size, to chase the speedy sardines far out at sea. They'd bought the large-meshed sardine nets from Peru. In turn, they'd sold off almost all of their used, densely meshed anchovy nets to the Peruvian fleets and to fleets south of Iquique.

"What a situation for a multimillion-dollar industry, caught suddenly essentially netless! At best, a few nets had been stored. However, the behavior of these new anchovies included a critical move forty to eighty miles offshore to deeper waters, where the older anchovy nets were too small and shallow," says Pepino laughing at impermanence. In sum, the sardine nets—highly maneuverable but incapable of containing the smaller anchovy—were useless. Thus, for several months Iquique's industrial fishing fleet sat at the wharf. It took about six months before enough mesh material could arrive in Iquique from Asian ports in Taiwan, Korea, and Japan.

"Why didn't they simply import from neighboring Peru?" I ask.

Pepino eyes me and says, "The fishing companies wouldn't order nets from Peru—they ran an unstable industry that couldn't be trusted to deliver, besides being higher priced." In all ways, the much-sought-after new nets were expensive and not easily acquired.

Pepino also reminds me of his own history. At this time, he was a captain seeking an anchovy net. He grins and continues, "Overall, friendship—*amistad*—is the key to our business. When I managed to wrangle a small, used net from a connection down south in Caldera, my prize sparked an unintentional bidding war. This is how I first flashed on the idea of quitting the sea to make profits on nets, to start a net company. I just waited a few more years."

Nets Reflect the Environmental Situation

Flip-flopping between anchovies and sardines became an industrial headache. From 1987 until the early 1990s, anchovy remained their key fish. But the seesaw struck again. Only this time, species pulsed back and forth in short intervals: one month might be only sardines; the next month only anchovies. A few months to a year of mostly one species, would be followed by another unpredictable shift, occurring at any time, seemingly out of the blue.

Musing, Pepino says, "The situation might have made net switching and net alterations a constant loss, if it weren't for net standardization. Net standardization started when everyone was stranded in 1987. Until then, a captain might dictate his net preference to his company's chief of nets. Yet, during repairs and alterations, spares to the captain's liking weren't necessarily available. The chiefs of nets began deciding to make all new nets uniformly deeper, and able to contain anchovies as well as sardines. Most drifted toward producing only anchovy nets because anchovies were appearing to be the predominant catch. And, these smaller-meshed nets did hold in sardines."

Pepino adds, "The logistics of constant net switching had driven everyone crazy. By 1993, the companies had grown uniform in their choice of nets. Net businesses, like mine, were in place servicing them." Climate and species shifts had contributed to a change in business practices and in net technology itself.

NEW TECHNOLOGY AND TECHNIQUES

The shipyards in Iquique continued to flourish with technological advances besides nets. The critical innovation was shifting to night fishing sardines far out. Out at the Pirate Beach Club, I'd been able to make an appointment with Ramon, a Pirate who's the fleet manager at one of the major industrial companies. Ramon had weathered the wild net sea changes, and begins telling me about concomitant ones in vessel types.

Walking along his industrial wharf, Ramon says, "Newer vessels of the Norwegian style—longer and narrower, with the bridge housed in the stern—were much more seaworthy to fish offshore in high swells and winds. To chase the sardines, we brought in and soon were adapting this vessel technology for our own purposes."

Ramón pauses by a rust-pocked hull being painted and continues the saga, saying, "Peru's fleets, like ours, experienced poor fishing as the 1982–1983 El Niño chased away the anchovy. At the company where I was working, Guanaye, we were able to purchase a few Peruvian vessels for a song that were in the larger class of 300 to 400 tons, wider and stable, and we just refitted them. We resold them in Peru when we moved on to vessels double that size up to 1,000 tons."

By 1989—a peak catch year in Region I—the bulk of the catch could be landed by about seventy mega-vessels from the six industrial companies, each with a fleet and shore plant: Guanaye, Iquique, Coloso, INDO, EPERVA, and Tarapaca (later replaced by Camanchaca). The largest among them was the aptly named "giant," Coloso, with twenty-three vessels based between Iquique and Arica, bringing in over 20 percent of the total catch.

"These days, fish-finding methods have become highly technological," continues Ramon. "Spotter planes that we first used in the 1970s became old hat by the end of the 1980s. By the mid-1990s, computers and satellites were becoming the norm in the tracking business. Cellular phones, for secrecy purposes, started in 1996. They really work well on the open ocean, where nothing blocks their transmission. I remember how until cellular phones, all the industrial vessel captains would eavesdrop on each other over the radio to figure out where to go. If they couldn't figure out where the good fishing was that way, they'd follow a vessel that had recently offloaded a lot of fish, knowing that captain would rerun to his good grounds."

Iquique's Four Major Industrial Companies during El Niño 1997–1998

During El Niño days in 1998, Pepino Hernandez's old building sits oddly quiet. One secretary remains within his historic building to answer the telephone or respond to someone ringing the ship's bell attached to the high wooden front door. She lets me in when I ring and points me to him at his desk in skylight beams shooting through empty high-ceilinged spaces. The contrast is stark to October a year ago when Iquique was on the brink of El Niño and Pepino was keeping over 130 contract workers busy. By March 1998, he had to lay off almost half the workers. By July, Pepino could only keep twenty-five net workers, dividing them between little jobs at two of the industrial giants, EPERVA and IGEMAR. Soon, EPERVA entirely canceled work on nets. This left Pepino reduced to a skeletal crew of six to equip the two new vessels IGEMAR had been caught building when the event hit.

Every giant was caught off guard. One of EPERVA's chiefs, a friend of a friend, meets me at Iquique's best bakery restaurant near the central plaza. I identify Manuel by his pressed checkered shirt with an air of business about him among the weekend throng, tables encircled by families enjoying large slices of decorated cake and fruit shakes.

Manuel starts in by saying, "None of us expected this year's Niño to be so hot for so long, or for the sea to be so unusually rough for so long. It has no precedent since the industry got rolling around 1960. Only the 1982–1983 Niño had a bite." We tackle thick cream cake powdered in chocolate as Manuel says, "We called the event of 1982–1983 El Niño *del siglo*—of the century—but he did no more than interrupt the industry for three months. Oh, and he was the harbinger of terrific sardine catches."

True, the El Niño of the century severely damaged inshore marine life, so artisanal fisheries suffered infinitely more than the Chilean industry. Interestingly, Peru suffered the opposite. There was nothing left to catch for the industry, however, Peruvian artisanal fishermen who changed gear to take advantage of the new species available benefited, like in the Paracas scallop fishery (Meltzoff et al. 2005).

I ask Manuel how the giant companies first reacted to Internet forecasts in early 1997 predicting another mega event starting in the summer of 1997–1998. Manuel says, "We'd developed a false sense of security. We didn't put much stock in the predictions, and didn't fear a big event. But when El Niño heated up in November 1997, we began to realize he had a very different character than the last one. We were surprised at becoming crippled."

Manuel pulls out a sheet from his jacket pocket and points out official catch numbers, "See how EPERVA's catch crashed by more than 90 percent in 1998?" I follow his finger across the page and stare at the paltry figures for January through October 1998. I wonder about industry acceptance of management's closed season. Manuel says, "Company chiefs easily agreed to this closed fishing season—*veda*—because we're ready to wait until April 1999 for tiny anchovies to grow and females to spawn. April is classically our peak season on the heels of the January-to-March closure."

Widespread Unemployment

The mass of Iquique's industrial crew sit unemployed. Iquique, with approximately one hundred industrial and semi-industrial vessels, each manned by about ten crew, equates to about a thousand seamen hanging around the docks. Shore workers, such as mechanics, electricians, welders, and those who offload are also workless.

I watch lots of men sitting in the bars, talking on street corners, and strolling around the industrial area. The women net contract workers who are laid off find jobs cleaning the town's toilets and floors. And on Pepino's tip, I meet some of the men who were laid off congregating behind the duty-free zone shops by the port. These former net workers pick up daily pay by loading trucks bound across the *altiplano* into Bolivia.

I also find and talk to lots of grounded fisheries folk by hopping on the shared cabs plying routes around town. Many fishing captains who've invested in taxis are personally driving their cabs for a living until their vessels are called back to sea. Everyone seems to share the optimistic prediction of renewed fishing in April 1999, after the veda.

The Event Put a Number of Semi-industrial Owners Out of Business

Vessel owners—*armadores*—in the Suzuki fleet have a strong sense of pride in operating their craft, similar to what I'd experienced in fieldwork in Galicia, Spain, among their semi-industrial fleet's *armadores*. In Iquique, fluctuating prices are ravaging this fleet. Many are forced to quit, their professional identity and pride crumbling around them.

By December 1997, temperatures by the coast were much hotter and the industry is left with neither sardines nor anchovies. High temperatures continued through April and even May 1998, paralyzing the industrial and semi-industrial fleet. By June, upwellings cool the coastal temperatures and a few anchovies start to appear. They swim extraordinarily close to the shoreline—too close for industrial nets or even the smaller semi-industrial versions.

Back out on the beach at the Pirate Club, Raul, another fisheries friend of Pepino, is facing the decision to sell his Suzuki vessel. He rues losing the title of armador. Raul says, "My upkeep and running costs are just too high, and I don't have a financial cushion like the industrial companies to hold me afloat." Operating on a single-vessel scale, Raul and his fellow armadores could not do what the industry does to limit double exposure to ecological and market risks. It's no secret to the armadores that the industry has been preparing in several ways to modulate impacts from the next El Niño.

However, it's the net contractors, such as Pepino, who could create new business opportunities based on industrial wishes to minimize exposure to legal responsibility to workers. The fewer full-time employees there are under contract, the more flexibility. Modernizing plants and fleets for maximum efficiency—as well as to meet environmental regulations—gives the industrial companies even more of a margin.

Although the industrial companies survive where semi-industrial individuals fail, the 1997–1998 El Niño event provokes extreme measures of industrial belt tightening, including firing down to a bare-bones staff of the most skilled. Company chiefs feel the heat from attending unusually high numbers of cost-cutting meetings.

Industrial Companies Experience El Niño

Angelini's EPERVA epitomizes industrial growth. So I explore labor changes at EPERVA, the industrial company started by the legendary entrepreneur Angelini. Angelini's seemingly intuitive business moves spearheaded Iquique's shift in the early 1960s to industrial fisheries with government support. His Angelini Group remains the most powerful in the fishing industry. His policies shifted toward workers to fit the political climate of the day, being able to fit in with both Allende and then Pinochet.

By the 1980s, at a peak in the industry, Angelini's EPERVA had merged with INDO, another industrial company. In the 1997–1998 event, they cut their Iquique workers almost in half. EPERVA's contract laborers dropped far more precipitously to zero. Of EPERVA's twenty-three vessels, sixteen kept fishing, with 11 crew each—making 180 crew at sea—plus 120 shoreside at the plant. EPERVA was able to keep the most vessels active during the event.

IGEMAR is the second company of the Angelini Group. It was started in 1991, absorbing Guanaye when Angelini became the major shareholder. IGEMAR carried out streamlining and firing from early on to modernize and become efficient. The company had a single, newly finished plant at the beginning of 1997, making it the leader of efficiency and environmental compliance—the new plant could process 180 tons of fish per hour, compared to 50 tons per hour in the old plant. It slashed labor and energy costs, including the use of water and gasoline.

During the event, IGEMAR was operating eight vessels—half the number of EPERVA—and ultimately cut back to four vessels at sea. The major companies do not foresee themselves employing the previous numbers again. Chiefs of EPERVA and IGEMAR in the Angelini Group, say that in the fray of this El Niño they never intend to return to high labor numbers and will continue seeking ways to operate more efficiently.

They don't fret over a collapse from overfishing or natural events. EPERVA predicts an eventual labor rebound to a maximum of 1,400 people, but intends to hold that line by increasing use of both contract labor companies and ever-more-efficient technology.

New technology includes updating vessels for the Angelini Group's fleet. They will be smaller, faster, and more maneuverable, like the 420-ton vessel class IGEMAR began building just before the El Niño collapse. One company chief passes on an important grain of wisdom, "When the fishing is great, almost any vessel works. But, when fish are few, we need faster, more efficient vessels to switch grounds and take advantage of any sudden situations that might arise."

Another key strategy that the major companies with multiple ports employ is to shift offloading ports depending on fishing grounds. For instance, EPERVA and Coloso maintain three plants in each of their three ports: Iquique, Arica (to the north), and Mejillones (to the south). Decisions to direct captains to different ports rely on fast vessels.

One of Angelini's naval engineer designer, José Miguel, claims that the ideal for adapting to grounds would be the even smaller and quicker 330-ton vessel with a speed of twelve knots. It would be far easier to maneuver among the many vessels all pulling nets. He describes to me a fishing trip he experienced, saying it was "a zoo . . . 105 vessels around us at sea all fishing at once, and there's a lot of bluff and near-misses as the big vessels try to maneuver without cutting each other's nets and slamming into anyone else. It's tense dragging nets at close range."

Another advantage of speed is that none of the vessels carries ice or has freezers when the final fish product is fishmeal, but the fresher the fish, the higher fishmeal quality, and ultimately the price. Four to six hours is the optimal storage period on board for prime fishmeal. The fish are fresh enough with this short time at sea for workers to take the fish home to eat.

In practice, forty-eight hours tends to be the maximum time at sea, reflecting cutting the edge between the maximum period a fish can remain in the *bodega*—fish holding area—given heat and decomposition. If longer, there is a real risk of fermentation. Wharf folklore reflects local fear and perhaps occasional truth, telling of the buildup of rotting fish gas that kills whoever is the unlucky one to fling open the hatch. Alternatively, the gas hides in pockets among the rotting fish down in the bodega and literally knocks a man dead who is offloading this putrid cargo. The moral is "Speed is of the essence." High-tech suction systems could replace manual labor, save suffering, and balance costs.

My next visit along the industrial row is to Coloso, where I concentrate on learning a bit about worker population profiles. Locals predominate as workers in industrial fisheries. Coloso is like IGEMAR and other companies in that roughly 60 percent of their industrial workers originally come from Iquique. This is a heavy local employment profile because Iquique's population overall is 70 percent migrants. Many came here in the 1970s and 1980s to serve in the army in the north, and many stayed on in all sectors, from Iquique's Duty-Free Trade Zone to fishing. With stiff layoffs, the industry is up to 95 percent of the workers coming from Iquique.

The chiefs of the companies—who are not from Iquique—insist there's no conscious plan to fire migrants before locals. When I go by the union offices, union leaders insist there's no coincidence and that the industry has a logical plan to prefer to fire local workers last because they'll be less likely to leave town given their local social security net. Moving elsewhere might have less appeal during periods of being laid off. Companies have an easier time rehiring folks who have stuck around. The current firing pattern lends credence to the union perspective, despite industrial leaders' overt intentions.

The actual numbers highlight the situation. What Coloso states is that, like other companies, it had to slash its crew and land-worker force by 50 percent. As of May 1998, plant workers, office workers, and special skills workers had dropped from 178 to 95. Of a sea crew of 106 on nine vessels, only the crews of four vessels were actively fishing. Coloso's contract workers by October 1998 were down to zero, contrasting April figures that showed the company employing 30 in nets, 16 in the plants, 10 in solid waste recovery, and 6 watchmen.

In the case of Camanchaca, 1985–1995 was the best time for this company. By 1985, they had updated technology, and by 1990–1992, their fleet was top of the line, with efficient, modern vessels. In 1990–1991, they had the foresight to diversify into salmon aquaculture in the south. While in Iquique, they maintain a fleet of eleven vessels, of which nine are fishing and two are on standby with the skilled portion of their crew: *patrón, motorista, contramaestre*, and *pangero* (captain, engineer, first mate, and net setter). But, the six seamen and the cook are temporarily laid off. Although it's advantageous to hold onto locals versus migrants when cutting back labor, temporary labor plays an increasingly vital role. In 1997, Camanchaca had 121 crew, with 90 percent on permanent contracts and 10 percent hired per trip of two to three days. By 1998 they still retained 103 crew, but only 60 percent were on permanent contracts; the rest were hired by the trip.

Proposal for Transferable Quotas

Rumors are that the 1991 Fisheries Law would be revamped in 1999. One plan is for transferable quotas. The idea is to encourage aggregation of companies, continuing Angelini's trend in Iquique when his group turned four companies into two by incorporating INDO and Guanaye into EPERVA and IGEMAR respectively, thus reducing duplication and the labor force.

One company hoping to benefit from transferable quotas is Camanchaca. The company is named after the north Chilean desert fog that rolls in off the sea at night and hangs about in the *pampa* desert, providing critical moisture that moistens an ecosystem of *algarrobo* trees and shy foxes. The company hopes to lead the way in efficiency as the newest and smallest company in Iquique, but the one whose family-run business is solely focused on fisheries investments.

The major owner is Don Fernando Valdez with 60 percent of the shares, who is still the active president and director of the company. His cousin, Don Francisco Sifuentes Carrera, owns 20 percent and is the *gerente general*—general manager. Camanchaca only arrived in the north in 1993, starting a plant and a fleet. Their goal is to come through this event, then buy up whatever the other companies are willing to sell next year, whether it be transferable quotas or vessels.

CONCLUSIONS

Political factors always play an important role in industry as governments adopt or discharge policies according to their own orientation and the changing value of resources. Iquique's fishing industry is a great case for observing these permutations, beginning in 1960 with the law to create industrial free-trade zones at the isolated extremes of an exaggeratedly long country. With the death of the important nitrate industry in the desert above Iquique, hoards of displaced miners descended from the desert, drawn by "the smell of the dollar." Jobless, they gladly converted to cannery labor for Iquique's fledgling factories. And artisanal fishermen upped their hunts, netting more fish to supply the new canneries.

After a score of years, when Chile's canneries could no longer compete with Peruvian and Asian companies, the industry switched to the even more odiferous production of fishmeal. Since the 1980s, fishmeal has remained the strong arm of Iquique's processing industry and Chile's main industrial fisheries.

Cycles of boom and bust have characterized Iquique's industrial fishing sector with El Niño events playing a fundamental role, alternating availability of their targets, anchovies and sardines. The first blow to Iquique's industry came from the 1964–1965 El Niño event. Anchovies disappeared, and a number of companies closed, leading to massive unemployment. Six companies emerged robust, able to persist during these tough years as stocks recuperated. Their survival strategy was twofold: shifting technology to new night fishing for sardines; and moving their fleets farther offshore where sardines (the new target) congregated. Post–El Niño, Iquique's fishing industry suffered a 50 percent reduction.

Soon the woes of the unemployed and investors were exacerbated by the pan–Latin American economic crisis of the early 1970s, compounded by Chile's politically induced economic turmoil. In 1970, socialist president

Allende responded to grave unemployment by nationalizing industrial fisheries. To create jobs, plants were forced to employ far more workers than necessary. A genuine fisheries crisis came from the powerful 1972–1973 El Niño event that collapsed anchovy fishing for the second time. Once again the industry was rescued by sardines in a similar seesawing of species, a lesson learned from the previous event. This time the collapse and shift occurred in conjunction with a major political event, Pinochet's 1973 coup.

Pinochet, hand in hand with the Chicago Boys' adventures in extreme capitalist reengineering of Chilean society, put emphasis on high production for export revenue. Giving out new credit to develop industrial fishing infrastructure, this sector started to grow again. In tandem with Pinochet's privatized export-driven policies, advances in freezer technology enabled burgeoning export from 1980 onward. Fisheries markets globalized and industrial fisheries prospered. Iquique's Duty-Free Trade Zone came into full swing during Pinochet's rule, boosting Region I's border economy, and translating into big money for business development including industrial fisheries.

To address what they viewed as out-of-control industrial fishing, the government consulted with university marine scientists in developing top-down regulatory measures. As a result, they imposed "limited entry," curtailing the number of vessels fishing for anchovy, horse mackerel, and sardines. The industry adapted by shifting to fewer but swifter vessels that doubled in size.

A number of highly skilled captains from the industrial and artisanal fleets, gambling on the change, bought up the industry's smaller and slower vessels. Helping the industry recycle their cast-offs, the government offered credit to this new group of vessel owners who formed the novel Suzuki semi-industrial fleet. They found themselves anchored within a gray legal area, defined as neither artisanal nor industrial and able to fish in either zone.

Environmental change in marine ecology alters how people practice business. With the event of the century—the mega 1982–1983 El Niño—net technology became critical during the classic seesawing between anchovies and sardines. Shortly after the event, the larger sardines appeared in enormous schools, replacing the abundant anchovies. Over the next five years, the wonderful wealth of sardines continued to enrich the industry and its workers. To increase their catch efficiency, almost all captains shifted exclusively to sardine nets with their wider mesh, thus having quicker haul times using less fuel.

Then the industry was caught off guard by the 1987 El Niño's stock seesaw. As sardines plummeted and anchovies soared, it was too late to amend the error of having switched completely to sardine nets. Months passed before enough mesh material could arrive to re-create nets able to contain the smaller anchovies. For three years anchovy stayed the key.

Everyone wished to avoid repeating the costly mistake of being caught off guard with the wrong net. Captains started carrying anchovy nets almost exclusively. Although their smaller mesh is slower to pull, anchovy nets hold in the larger sardines. When sardines and anchovies flipped again, this time it led to net standardization: making all new nets uniformly deeper, and able to contain both species.

Ever since, the entire industrial fishing sector revolves around the use, design, repair, and trade of nets. Net-maker leaders decide the final net design. Most of them have learned their trade from experience, not schooling. Lately, some net leaders are fisheries engineer graduates with university specialization, just like a growing number of captains who are emerging with degrees.

The fleet has also discovered a critical fishing innovation: sardine night-fishing far offshore. Iquique shipyards have complied by building vessels along the "Norwegian style" that is more seaworthy and incorporates high-tech fish-finding technology.

At this point another giant event arrives—the 1997–1998 El Niño. Although industrial fleet managers were reading competing websites that vie to interpret the NOAA climate predictions and were watching for signs at sea that they trusted, none of them are prepared. All major fisheries companies are caught off guard. Their expectations for a seesaw are dashed. Instead, no fishmeal species can be caught. Nobody predicted this El Niño to be so hot and to last so long, or for the sea to be so unusually rough. Industrial companies could survive this individual El Niño, but many semi-industrial vessel owners have had to fold. The event provoked extreme measures of industrial belt tightening. The industry has fired down to a bare-bones staff of the most highly skilled, and has mechanized where it can. This limits exposure to labor needs and contractual demands, as in any big business. The new bust cycle does not last long. But the shift to labor subcontracting is the legacy. Iquique fishmeal plants have also been legally required to upgrade operations, renovating to meet stricter codes for clean air and water emissions.

EPILOGUE: IMPLICATION OF EL NIÑO PREDICTIONS

As part of their approach to modernity, Camanchaca and the other industrial companies' fleet managers use daily sea surface temperatures that affect fish movement. For instance, temperature-sensitive anchovies seek the colder 12°–13°C waters, are okay at 14°–15°C, but die at 16°C. Thus, having temperature information is an advantage to a responsive, fast fleet. Companies value NOAA's daily sea surface temperatures and satellite services that monitor daily surface water temperatures. When deciding where to deploy the vessels, fleet managers on land increasingly supplement their customary reliance on skilled captains' intuition.

El Niño predictions for fisheries, however, are in nascent stages of reliability. Climate modeling is subscribed to in Chile, as in Peru and Ecuador, but it is primarily university and government biologists who are playing with multiple and contradictory websites and data, interpreting it for the industrial sector, who have the business budgets for access and scientific interpreters. Fisheries clients hope that accuracy of forecasts will improve to the point where they can integrate El Niño predictions into proactive planning. They hope for forecast advances so they might keep from making new investments at precisely the wrong moment. But so far, industrial fisheries, like the artisanal fishermen, are unwilling to risk acting more than ninety days in advance, and need to see the in-the-water changes for themselves first.

All those who harvest the sea learn that El Niño events are individual in character, each coming with permutations critical in fisheries: duration, intensity, and location on- or offshore. While one event may stay offshore or turn out to be relatively tame and short, another may sweep onshore, bringing stormy seas, and may last a long time. The great 1997–1998 El Niño reminded everyone in industrial fishing not even to expect historical patterns—like the safety net of the sardine/anchovy seesaw.

A Camanchaca chief notes that nobody within the industry planned on the 1997–1998 El Niño being so strong, or strong in those precise ways. If they had advanced predictions, they could have been able to regulate spending better and to begin economizing a half year in anticipation—also the time it takes to import materials. "Such forecast precision," the chief sighs, "is still in the realm of science fiction."

Coastal communities dependent on fishing-related industries realize unending ecological change will continue bringing some times of poor supply. In response, they are consolidating fishmeal factories and contracting labor to limit exposure to costs. A few have even decided to invest in neighboring southern Peruvian fishmeal factories, further hedging their supply bets. The industry also insures their survival by ensuring flexibility of its fleet. Vessels of the major Iquique companies have long-established government permits to fish throughout the Chilean fishing zones, to be able to "migrate" to follow stocks.

Industrial fisheries, from fishmeal to merluza, fund biological research and use its political weight to keep the total allowable catch constant, ignoring El Niño's impact on recruitment. This "liberal" approach to management, based on current fish mortality, recruitment, and abundance, ensures the industry's flexibility to fish unencumbered by more stringent, "conservative" management regimes. The fishmeal industry feels this short-term management style is to their advantage, lending flexibility to cope with the inevitable

El Niños dealt by nature and experienced more profoundly as one moves north toward the equator.

LETTER: NORTH OF IQUIQUE FROM THE PISAGUA COAST UP INTO THE HIGH DESERT

Hey Suzie,

The only palms here are well-watered ones. How are your fall leaves and kids' costumes? Here are some more stories from Pisagua and the altiplano toward Bolivia, where I went for Day of the Dead.

I rent a pickup truck for a week, this time wisely tuned for high altitude, and return to Pisagua. This time Pisagua was just locals diving for their octopus and ruing the lack of loco, the abalone-type shellfish that they illegally smuggle to the Peruvian market at any size. This year loco are just too small and seas too rough by the rocks where they dive. So men hang out a lot and like to pass the time talking. Last year I'd visited their sea lion nursery with Checho (wife Aurora, great name isn't it?) and taken his picture rowing standing up, crossing oars and pulling in the foreword position. We were coasting up over the swells near the rock islets draped in sea lions. A few females slipped off the rocks into the blue and wove around us, pushing among themselves over space like kids in the backseat.

This year Checho takes me back to the sea lions in the early morning light. We see mostly the young and some full-grown females. The ballet surrounds us, leaping and gliding in their sea lion element. The only bulls who survived El Niño are the smaller young ones. By the wharf, El Gortito Regalon—The Fat Pampered One—sits at his table in front of a yellow shack where he runs two freezers for dealing in shellfish—mariscos. He sells sea lion oil as medicine but especially as an aphrodisiac. He's one of ten locals who grill the sea lion steak for a potent meal. The supplier of town news, Gordito is a transplant from elsewhere, like most of the other forty men in town. Too obese to move, he banters and barters his living, selling shellfish up in the altiplano and to the weekend winter tourists from Iquique and the summer crowd who come for the beach and the distinctly desert sun.

Everyone in town awaits the completion of the wider road descending the cliff, and the arrival of electricity. Both are expected by next March. No more candlelight dinners but more hours of TV; less generator rumble in the evening and more "convenience," say the women. About ten of the divers have shifted to road work this El Niño year, earning steady construction pay. Divers from Iquique and Arica come to Pisagua this year as usual to seek the abundant octopus.

In an ironic political twist, Rafael, who was a cameraman for Allende, and his wife, who taught French in Santiago, have moved to Pisagua to find paradise. They arrived two years ago and bought a piece of land with an open view of the bay by a huge boulder, up above the police post. Rafael had filmed President Allende visiting Pablo Neruda's famous beach house at Playa Negra, and this had become his house ideal: a wooden structure like a ship's prow overlooking the sea with great rocks as natural sculpture.

Rafael has formed and become the president of the *Junta de Vecinos*—Neighborhood Association—and his wife has made a cultural group connected to the old theater, a decorative wooden gem where Caruso sang and Sarah Bernhardt performed. Rafael is writing a script for TV about the coup they survived, and his wife paints flowers in this desertscape. Interfacing with the government, Rafael has been able to organize some women, but not the men, to change the town.

While I was there, the national senator of President Frei's party, a woman lawyer from Iquique, came with a haircutting team and a pair of clowns, plus two singers, to talk to the people about her development plan for them. The younger men and the women and girls took turns sitting in a circle in the plaza by the canon facing the sea getting haircuts. Then everyone went to the old theater. In a place of little entertainment, all ages laughed at the clown act with a lineup of little girls in a dance competition. A plump sexy woman sang, and last a boy replete in a Mexican silver-threaded hat (who has appeared on popular Miami TV) brought down the house with his vibrating voice.

In the plan, tourism is the linchpin. Forty families will be given new houses higher up the hill, while the coastal area, where some have made homes will be razed to become a tourist playground. There's also a study underway by the government to determine who owns the town's delightful dilapidated old wooden houses, if anyone. Who needs a solid roof or walls when it never rains or snows? The squatters will have to give way to vacation-home owners, who will renovate.

The owner of the town's hotel in the former Pisagua prison is a conservative from Arica. He says he would only sell out if the price was high. The political Left who come on annual pilgrimage to Pisagua's mass grave hate the hotel being in a place of torment. Instead, they want the prison to become a museum-monument in honor of those imprisoned and killed by Pinochet. The owner steadfastly notes he spent considerable money over seven years renovating the wreck into a decent hotel in hopes of a tourist boom.

Climbing back out of the gorge, I'm giving a lift to Jose, who usually drives his aunt's truck to Iquique to sell Pisagua mariscos. He asks to drive, and I'm only too happy to hand him my keys. In the backseat are the diver's family—the mother with babe in arms and her small son, and the grandfather, who is also a diver. They're headed to the highway to catch the bus north to Arica where the mother works in a hospital. Families are often divided for work. They point out the old railroad line cutting across the sand hills above us. Rail was the lifeline of Pisagua's nitrate port in the last century, abandoned like all the nitrate mine settlements along the way marked by earth-returning-to-earth walls, and a few crosses. At the turn up into the mountains, Jose goes to find the Iquique bus, and I head up to rarified altitudes.

Three days later I'm sleeping at 4,300 meters ringed by volcanoes at Salar de Surire, a national reserve for vicuna, a wild camelid almost killed off for its incredibly fine silky fur. Up here they are plentiful and the state is the guardian. At the park refuge where I sleep, there are two Chilean veterinarians and an Oxford wildlife biologist carrying out a four-month study on the impact of shearing on vicuna. Because it takes a vicuna a year to regrow its fur, the park service wants to know whether to continue to allow this local practice or if vicunas' health and survival is in jeopardy in the bitter cold. *Ñandus* (small emu/ostrich/rhea type bird) run along the salt flats on powerful legs. I see a female who stays alone, and a male with his gaggle of chicks doing their best to keep up on short legs. I'm pretty dusty from the whole morning's drive, which turned out to be more orienteering to pick the right track northward along the Bolivian frontier. But I find the thermal bath on the salar. On one side, it's boiling, bubbling, steaming. Downstream in pools farther away, I locate bearable water at Japanese *furo* bath temperatures and gradually settle down into the mud soufflé, water up to my neck sitting. My silver ring quickly blackens from the high sulfur as I play in the creamy, gray dense mud. I collect a few handfuls as it's exactly the thermal mud I wash with at home. Meanwhile, a flamingo joins me a few meters away in a slightly cooler area, sieving her lunch from the mudflat waters.

The next day is Day of the Dead (*Dia de los Disfuntos*), 2 November, and I return to Chusmiza, via a geyser that creates an ice mound, with a full family of Aymara Indians in tow, riding in the back of my rented pickup. The grandmother is in front with me pointing out the best pastures for llama and alpaca, and naming each volcano and hill. I reach Chusmiza, the thermal spring pueblo where I'd met people last year decorating their loved ones' graves. I'm in time for the fumarole pan pipes and drum beat. When the drummer gets too drunk from the boxed wine offered by each family, the music group recruits another guy to keep time.

Now, two days later, I'm back on the coast interviewing people in Los Verdes, the fishing caleta just south of Iquique. Because this year has been extremely hard, a number of wives in the rural caletas have started selling their husband's catch, where before they focused on being housewives. Necessity.

Better write field notes, my necessity, but I wanted to send you many thoughts and good wishes.

Hugs to Jessy, Ben, Spence, and Ali and tell them about the ñandus and flamingoes for me . . . OXOX Sarah

Chapter Seven

Sea Wolves

Artisanal Fishermen of the Rural and Urban Caletas

SEA LIONS FEASTING OFF FISH GUTS FROM THE WHARF

Sea lions float around Iquique's only two urban caletas, Riquelme and Cavancha, for daily fish scraps. At Caleta Riquelme, I'm watching them lolling on the rocky shoreline, leaning into each other, embracing, belching fishy breath. They lunge at audacious gulls or pelicans trying for a morsel. The imposing bull, a veritable torpedo in water, lumbers through the shallows toward a fisherman tossing down fish guts from the wharf above. He catches and casually tosses back bits into his mouth, whiskers quivering, eyes brown pools. With his massive neck coated in thick honey fur, to me he resembles a lion more than the sea wolf of his Spanish name—*lobo del mar*.

ETHNOGRAPHIC STORY AND EMERGENCE OF GLOBAL THEMES

Voices of Old Sea Wolves in the Urban Caletas

"Sea lions never used to venture onshore to beg for scraps like this, like circus seals," says Don Vega as we stand side by side on Caleta Riquelme's high wharf (see figure 1.2). Iquique's two urban caletas, Cavancha and Riquelme, have existed since before the establishment of the fishing industry (figure 7.1).

Don Vega is a senior fisherman out of a Hemingway novel, forceful, tousled, and lined, a prominent captain who is successful at finding schools of fish. He points down upon the sea lions, again noting how this undignified

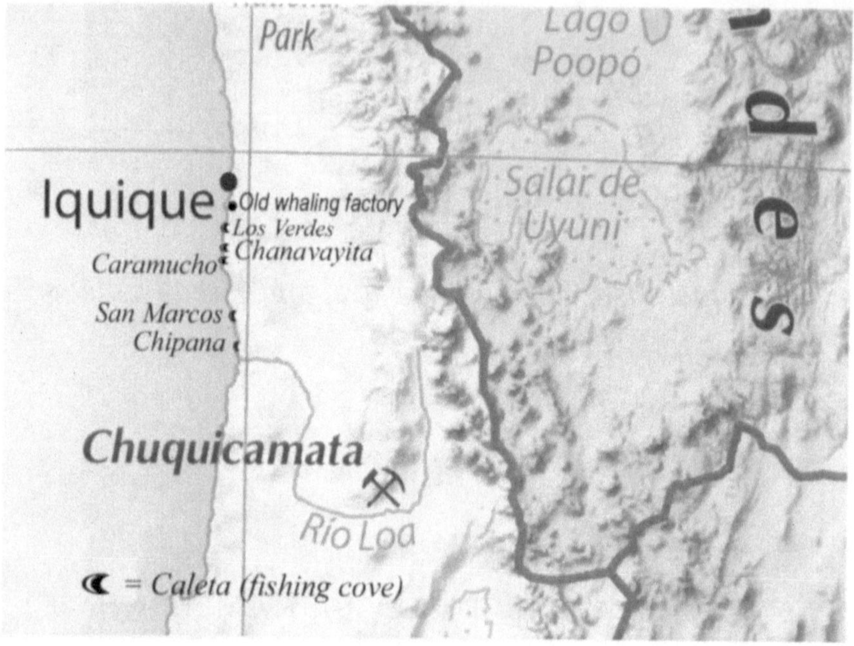

Figure 7.1. Rural caletas south of Iquique map. Adapted from base map from https://www.cia.gov/library/publications/the-world-factbook/.

phenomenon only started within the last decade. "Iquique's industrial fishing trawlers are to blame, swiping all the fish. They set their giant seine nets within five miles of shore and sweep up the little baitfish that draw in our catch. Without those baitfish, sea lions are hungry like us artisanal fishermen. It's not fair! This is why sea lions are now bumming around the caleta for scraps."

His apt metaphor between lobos and artisanal fishermen is clear, and I feel his frustration. Don Vega is the elected leader of Caleta Riquelme, the members voting for him as their caleta's mayor—*alcalde de mar*—their public face, the one who daily issues permits for every fishing trip. With the other artisanal leaders in Region I, he's embroiled in the region's political battles against the industrial fleet to maintain and enforce artisanal territorial fishing rights to the first five miles from shore.

This five-mile zone is legally designated as artisanal fishing grounds, but the industrial fleet routinely ducks inside to fish on the prolific narrow coastal shelf. There is exactly the same frustration in Peru, same laws, same conflicts between the artisanal and industrial sectors.

Don Vega's face reddens and his mouth works. He says, "It's our baitfish nursery being wiped out by the industrial leviathan! With them on our grounds, we can no longer catch quantities of fish that locals buy for dinner at our caleta market. Bluefish has been a staple, but because this fish hunts baitfish on the shelf like the industry, they also have become scarcer. We're being forced to accept scraps from the industry." He's punning since the industry literally gives the caleta leftover scraps of netting to rework into their net shapes.

Two birds fly overhead heading north. Standing on the dock at the other urban caleta, Caleta Cavancha, Antonio pauses to tell me that means good weather ahead. Antonio is highly regarded by his peers as a fine captain having vast experience, having fished from the south in Valdivia where the great Humboldt Current—now called the Peruvian Current—approaches the coast, all the way up to El Callao in Lima, Peru.

Antonio's talking about his childhood. Born in 1935 into a fishing family right here on the little peninsula of Cavancha, he says, "I remember the old days when Cavancha was a colony of fishermen, and we'd take buses northbound into Iquique that only ran on Saturday and Sunday. Back then we had no light. Water was delivered in trucks, and only in 1945 did the town finally run us a water line. My mother and most women grew little gardens to augment our daily seafood supper. Women went by burro carts up into town to sell their husbands' catches.

"Artisanal fishing typically entailed leaving home for months at a time, migrating to uninhabited caleta grounds to find fish," Antonio says. "My mother didn't like how this took us away from home for maybe a quarter or half year at a time. But we'd have to move around in order to survive. We'd set up camps along the coast, net-fishing or diving for the town markets. It was a *puro hombre*—male-only—lifestyle, rough and a bit lonely, but it let us catch whatever was available that had a price. We could exist given the ocean's shifting bounty and her constant changes."

I ask him about formal education, and Antonio says, "Most of us finished four grades of school before going to work at sea. I set out full time at fifteen years old with my Tío Humberto, a great captain. I crewed and learned from him until he switched over to the industrial fleet as a *patrón de pesca*—fishing vessel captain. Lots of the best artisanal captains have crossed back and forth between sectors, trying to foresee which fleet could earn them the most under current conditions." As Iquique's industrial fleet expanded and contracted operations during climate events, a number of artisanal fishermen flowed in and out as captains and crew, jockeying for a level of income.

I'm watching Antonio prepare for a ten-day fishing trip. Besides being captain of his vessel, he's proud of being *el lanceador*—the one who balances on the extended prow of the boat, hurling a lance into a billfish as it breaks the surface. On the dock, Antonio plants his feet wide, throwing one arm back high to demonstrate the stance. Weighing his lance in hand, poised and still, his concentration is palpable. I tell him how I find it amazing that this type of hunting is still practiced. He nods and adds, "No rifles, we don't use them. Rifles just aren't as accurate at sea tossing and rolling on the swells."

"What's the top issue for you as el lanceador?" I ask.

"The problem today is sighting any type of billfish in the first place! Marlin and swordfish, they're becoming scarcer and scarcer. During my twenty years hunting, numbers have crashed. For every one hundred billfish I'd spot before, I can find less than five," says Antonio. He continues with a sigh, "To make a living now, I have to rely on gill netting other species." His romantic era of being the oceangoing hunter and spearing billfish is fading fast.

Once Antonio is back from his trip, we meet near Caleta Cavancha in a café to continue discussing artisanal fisheries, billfish fishing in particular. He talks gently, almost lovingly, about the billfish abundance of the old days. "For the big fiestas in June or July, we could bring in fifteen tons of fresh billfish. During the four-month season in waters just one hour out from Iquique, I'd expect easily to harpoon eighty billfish. This last trip I only spotted two and landed one." Shifting to a rueful tone, he continues, "I have to hunt far afield to the south to make a trip worthwhile. But, if I see a handful of billfish this whole season, I'll be lucky."

I ask when his last good year was. "Perhaps 1973 was really the last year we still found plentiful billfish around Iquique. I think we can blame 20 percent of the loss of billfish on shifts in the currents, the calling cards of El Niño." Antonio continues with a penetrating look. Pointing in the direction of the plants nestled at the north end of town, he intones, "But the other 80 percent of the billfish disappeared because of industrial contamination—all the noise, fuel, blood—and tons of juveniles being swept up in their giant nets, wasted as by-catch!"

Industrial Power and the Bureaucratization of Chilean Fisheries

Iquique, with an approximate population of 150,000, is one of Chile's most vital industrial fishing ports, and has always been Chile's top port for the pelagics—fish that live in the upper layers of the ocean, either inshore or offshore—converted into fishmeal, especially anchovies and sardines. The city is home to numerous fishmeal and fish oil plants, that remarkably accounted for over 11 percent of Chile's total export earnings in 1989 (chapter 6). This nationally important industry carries considerable economic and political clout when it comes to competing with the artisanal sector.

To identify boats on the water, each port has a strictly enforced color code. The bright colors on the artisanal boats signify home ports. For instance, in Region I, Iquique is orange; Arica is yellow; Tocopilla is blue; and

Antofagasta is red. Fishermen must also be registered with their corresponding caletas, and the National Fisheries Service keeps up-to-date detailed books on each caleta's operations.

For the Iquique area, sharks, octopus, swordfish, and *corvina*—sea bass—are frequently listed in the caleta catch logs. But the most common fish, one that appears constantly in the logs, is bluefish—*cojinova.* Cojinova is still the local staple, not destined for the fishmeal plants. It's sold fresh or fried for lunch and dinner on the beaches, at the urban caleta markets, and in the secondary fresh fish markets.

Political Clout at All Levels

Socialist Cooperative Era, Heyday of Urban Caleta Power

During the 1960s, socialist influence at the national level was building. Both of Iquique's urban caletas, encouraged by changing policies, reestablished themselves as fishing cooperatives to qualify for government infrastructure development. This cooperative era increased caleta powers. Caletas were expanding and modernizing as important spaces where artisanal fishermen congregated and organized socially and politically. The co-ops provided docking space and storage for gear, and their own cold stores to be able to market catch to the best advantage.

However, the fishing co-ops failed. Besides collapsing under Pinochet, some fishermen blame mismanagement in the form of corrupt leadership, poor pricing tactics, and lack of training. A common co-op scam had been to file worker training paperwork in order to receive the tax deduction, but then skip the training process. Intermediaries took advantage of the co-ops, intervening in the marketing aspect because money could be made. They disrupted the ideal of co-ops: paying the same price to all fishermen, employing cold stores, and timing the sale of their own product to a wider regional market.

Artisanal Voting Block Starts to Matter

Once again, with the rise of the newly democratic government in 1990, the political voice and votes of artisanal fishermen began to matter. Artisanal fishermen lobbying as a block of 65,000 in Chile impact proposed laws. Their voting block is considered politically important by a government that's no longer socialist or military. When the 1991 Fisheries Law came into force, it was a major political milestone. The government granted the long-standing artisanal request for a five-mile zone—an exclusive artisanal fishing zone that extends out from shore. Immediately this threw the artisanal and the industrial fleets into conflict. This is particularly true in Iquique, Region I, because here the coastal shelf narrowly hugs the shore and Humboldt upwellings along this wall lure great schools near shore.

Fishermen Divided into Urban and Rural Caletas

Artisanal fishermen are divided into two very distinct groups: those organized into urban caletas and those making new settlements in rural caletas. Around Coquimbo and Tongoy (chapters 4 and 5), national policy makers first took aim to end fishermen's peripatetic migration practices and anchor them into a chosen caleta. Management Areas are intended as the next artisanal management phase, following on the heels of the establishment of permanent settlements. From Patagonia (Islas Galas, chapter 3) up through Peru (Meltzoff et al. 2005), crude shelters and tents—*rucos* and *carpas*—are being replaced with simple cinderblock or wooden houses.

Women and children have joined the men, ending the once all-male camping scene in temporarily inhabited caletas, replacing it with a modicum of family life. Driving across rocky sand tracks off Iquique's southern coastal road, I want to get to know the relatively new family lifestyle of the rural caletas.

Linked with policies for forming Management Areas, the government hopes to motivate migrant fishermen to convert all the temporary fishing camps nestled in the best fishing coves along the coast into these "permanent" enclaves. These caletas, ranging from fourteen to forty households, are set back off the only road running southward from Iquique to the airport and onward toward Antofagasta. Electricity is only available to those with a generator. No rain falls on these salt-sprayed shores; the Iquique government trucks bring in water once a week for the settlers.

Pivotal distinctions between the two urban caletas and their newer rural counterparts lie in the fishermen's places of origin, fishing techniques, and target species. Rural caletas that are no longer campgrounds are populated predominantly by migrant divers who brought hookah technology—divers descend with long hoses connected to a surface air compressor in a boat overhead—with them from the Coquimbo region or farther south. They maintain family ties and active social networks that allow them to return home during hard times to fish with their friends and relatives, despite the new illegality.

Until 1973–1975, hookah diving did not exist in Iquique. Urban caleta fishermen worked from either small open boats called *botes* or *faluchos*—the largest artisanal boat with inboard engine and cabin. Some breath-hold divers walked and waded along the coast, simply swimming out from shore carrying a net bag at their waist. By 1975, a cheap new Chilean technology was enabling many more fishermen to buy air compressors and shift to hookah diving (figure 7.2). During and after the huge 1982–1983 El Niño, hookah divers were able to descend far deeper, going down to fifteen meters to exploit lucrative unharvested areas in cooler water.

Rural caleta fishermen with hookah rigs target shellfish, especially loco, octopus, sea urchins, limpets, and clams, and some seaweeds. Meanwhile, the urban caleta fishermen set out in their seven-meter boats with nets, longlines, and handlines, going a maximum of one mile offshore, usually at twenty-five-meter depths. They normally land fish such as cojinova, the popular fried favorite.

During the warmer waters of intense El Niño events, urban caleta fishermen can catch more tropical species that swim into their nearshore waters, so fishing actually improves. But no artisanal fisherman is always blessed with labor and resource opportunities during El Niño. The cold waters of the 1999–2000 La Niña event drove many stocks to depths of eighty meters, so both artisanal fishing and diving deteriorated.

In the mid-1990s, an urban Iquique association of breath-hold divers—*buzos a pulmón*—without an urban caleta to call home base begin angling for their own rural caleta, complete with land and marine tenure. In town, they have neither boats nor a shore presence. They simply wade out and swim underwater, free diving to collect species close to shore, requiring no boats and minimal investment in gear. This association has regrouped itself to

Figure 7.2. Climbing back into the boat from hookah hose diving. Courtesy of the author.

become an official rural caleta, colonizing the uninhabited Caleta Caramucho, and claiming the rich marine grounds in front as their Management Area.

From the perspective of residents from the other rural caletas, these city-dweller divers moving into Caramucho are the interlopers, reaping the best marine resources while many of their families remain in town having the advantages of an urban lifestyle: schools, running water, on-demand electricity, supermarkets, cinema, Internet, and an overall exciting social life.

Rich Northern Grounds and the Contested Five-Mile Zone

Iquique epitomizes fabulous fisheries in Chile. The continental shelf narrows sharply in the north, creating a dynamic of cold upwellings rising up against the coastal wall that plunges to over two thousand meters deep within the first five miles of shore. This causes incredibly rich nearshore fishing grounds, some of the most productive in the world, a trait shared with southern Peru's contiguous waters.

Thus, Iquique is a striking place to witness the interrelations between industrial and artisanal fishermen. Iquique's history is synonymous with Chile's development of pelagic industrial fisheries in the early 1960s and the shift to exporting fresh seafood. Iquique's industry is also linked to the Peruvian fishmeal industry through investments in processing and vessels. Artisanal and industrial fishing sectors have coexisted in urban and rural settings, functioning at various stages of cohesion under a variety of governmental and cooperative fisheries organizations, and enduring numerous environmental, economic, and political climate changes. Both sectors have grown highly organized, managed, and competitive.

Whereas most artisanal fishermen consider government interest in their affairs a "headache," there are some financial benefits in the form of commercial incentives, training, and infrastructure via government aid. The artisanal sector has also won rights to the five-mile zone. This comes with caveats, and with continued expansion of Iquique's industrial fisheries.

The artisanal sector considers the five-mile rule a victory, but they feel outfoxed by a gaping loophole: if the government decides artisanal fishermen are "under-exploiting" coastal stocks, then industrial fleets can be granted permission to sweep inside. The loophole agrees that the first five miles belong to the artisanal sector, unless they harm the national economy, for example, not harvesting lucrative fishmeal species. Using this argument, the industry is allowed in, except for the very first mile (two miles if directly in front of a caleta). These extremely desirable first five miles remain a sharply contested political prize sought ardently by both the artisanal and industrial sectors, a situation mirrored in Peru.

Iquique's rural caleta folk experience the direct resource competition with industrial vessels that sweep into their nearshore waters, albeit illegally. Given the narrow continental shelf concentrating rich marine life, it's hard to resist dragging the area with massive trawl nets. They damage the benthic (bottom) ecology where caleta octopuses hide and the divers hunt. In the caleta, the law seems like a paper tiger as they see their grounds being churned to rubble by the drag nets heavily weighted along the bottom. The industrial fleet can ignore regulations because there's no real enforcement. Even the sacred literal first mile, beyond the reach of any legal loophole, is unprotected. Frustrated, caleta fishermen often speak of watching helplessly from the beach or from their small boats as vessels sweep in close to shore pulling nets. Thus, the caleta dwellers experience the boon of the five-mile zone as a sham. Industrial captains know the law, but profit is irresistible; conflict between sectors continues.

Alberto walks into an Iquique hardware store that stocks outboard motor parts and begins exchanging news with a few other rural caleta fishermen. We're introduced as conversation swirls about. Several men are complaining about the tiny enforcement base being in Iquique, too impossibly far away to discover any illicit activity and arrest the unwelcome trawlers. Alberto says, "Last week we sat watching two vessels not a mile from our house, nets set in the water under way, but what could we do? You know how it takes forever to get a patrol down to make an arrest. By the time officers arrive, the show is over. There's nothing left to see."

As shellfish divers, Alberto and his peers witness the aftermath, the loss of precious habitat. Underwater, they see the devastation left behind by the industrial nets that rumble across the ocean floor to entrap their target, bait-fish schooling in immense balls. Alberto shakes his head and says, "They rip apart the octopus lairs where we dive, and crush and scatter the sea life."

Shaking fists hasn't worked against the looming political and economic presence of the industry. The widespread artisanal sentiment is that the five-mile zone is bogus; there's meager enforcement even against vessels dragging right near the beach.

Balancing between Sectors: The "Suzuki" Semi-industrial Fleet

Nicknamed "Suzuki," the semi-industrial fleet's vessels have enough power and range to purse seine. The fifty independently owned and operated Suzukis were double-dipping into artisanal and industrial fishing grounds, falling between the legal definitions. Neither artisanal nor industrial by classification, they held nebulous rights to fish anywhere they pleased. Yet they were harming the seabed with their bottom purse seining—*cerco de fondo*—thus competing with artisanal seiners for the same species of fish, particularly cojinova and *pejerrey*.

As with fishing booms, government efforts at regulation only arrived after the fact. In 1996, industrial and artisanal fishermen signed a two-year accord that's another milestone in their dialogue. The accord declares a moratorium on the Suzuki fleet, finally forcing each vessel to declare either artisanal or industrial status, then to abide by that set of policies. Consequently, the semi-industrial fleet has shrunk to only about twenty vessels, having to compete with the far more efficient and faster industrial vessels.

When the government finally forced the Suzuki to choose between artisanal and industrial status, artisanal spirits soared. They want the Suzuki fleet far away from their precious first mile. Soon, the colossal El Niño event in 1997–1998 changed the game. In artisanal fishing circles I hear fishermen say, "It's the act of God, who knew no other way to force the big vessels to stop their coastal raiding."

Modern Iquique and the Urban Artisanal Caletas

In the late 1990s, the new airport of Iquique opened to fanfare, including a visit from Chile's president. Stepping off the plane after a year away, I admire its modern aesthetics and efficiency. The whole city seems to radiate a sense of progress and money. All along Baquedano, the main street leading to the central plaza, twin rows of freshly painted wooden, multistory, balconied beauties—national monuments from the late 1880s—reflect against the deep blue sky. Inadvertently, Iquique has held on to quite a few of these architectural gems by skipping the last round of mining development that took place above Antofagasta to the south. Iquique is at the heart of the current cycle of international mining activity up in its interior.

The wealthy are gaining cachet by renovating these historic buildings, as well as developing Iquique into a beach resort replete with casinos. High-rise condos and hotels with ocean views are shooting up, as are sleek, low-slung, Mediterranean-style, stucco houses climbing the hillsides at the southern end of town, enclave of mining executives and other high rollers. The new building lends a profusion of color that reflects off the desertscape in vivid rusts and salmons, deep turquoise and olives.

By lunchtime, I'm sitting at the little sidewalk table of my favorite restaurant in front of the central plaza, enjoying their daily fresh fish and soup menu. Along comes an exotic contraption labeled MADVAC, a mechanical street cleaner in elephant drag. Its trunk is swinging back and forth, sucking up street litter. I burst out laughing at this playfully mad modern pachyderm, who for me epitomizes sweeping change in Iquique.

Caleta Cavancha in the New Moneyed Stratosphere

Caleta Cavancha's once-isolated low peninsula rests between beaches well south of the central plaza of the historic downtown. Caleta Riquelme—the other urban caleta—sits on the sea a handful of blocks from the plaza, near the industrial wharfs and the duty-free zone near the sea cliffs that bookend Iquique (see figure 7.1). Iquique physically can only spread southward. So Cavancha finds itself caught in the midst of intense coastal development, land bought up and converted into high rises and other pricey properties.

Antonio, as one of Cavancha's old-time leaders, describes how their caleta faces land-development pressure similar to what I observed in the Florida Keys (Meltzoff 1997). There the original occupants are the backyard commercial lobster fishermen, who must contend with second-home buyers, and retirees moving in and trying to zone them out of their neighborhoods. In Antonio's boyhood, the peninsula had been a fishing community, long before they were bought out. Most fishing families have scattered inland into the town. They just maintain the caleta with its wharf where they can keep their boats and gear by the water.

Antonio and I wander around Cavancha's sliver of beach. Running his hands along the gunnels of wooden boats pulled up onshore, he analyzes recent developments. Antonio says, "The rich have absorbed the last land of this area through political power. Even Iquique's former mayor now has his house where a fishing family once lived. And see those cranes crowding around outside the caleta? More condos are under way."

Cavancha is now the upscale play hub of the city, and thus of all Region I. Club Kayak started one year ago, at the end of the elongated wharf in a makeshift container shed/boathouse. Club Nautico, sporting expensive curved architecture and blue awnings, rests to one side of Cavancha's land, and Club Yacht to the other.

Antonio throws his arms wide in both directions, "And look! Club Nautico and Club Yacht are upon us! Club Nautico's owners want to expand into Caleta Cavancha. Luckily, first they must address the city's rules against cutting off public beach access, and do something to make up for commandeering the existing beach. Their lawyers are busy repairing mistakes. Club Nautico has come out in the press promising the public environmentally sound development practices, and vowing to reopen beach access soon. That's bad news for us, because then they will refocus their energy on moving into our little remaining space. And they call the caleta an 'imposition!'" Antonio snorts.

The working-class landscape of the caleta with all its repair of nets and boats is labeled "ugly" by upscale newcomers who want the whole desirable place for themselves. As in the Florida Keys (Meltzoff 1997), people who invest savings in pursuit of fashionable pleasures desire neighbors with similar class aesthetics. The original fishermen, seen as dirty, noisy, and annoying, don't measure up.

Antonio and I walk out along Cavancha's wharf, pausing by a woman in a flowered dress. She's Anita, a captain's wife, born and raised here. She's leaning on the pilings by one of the boats at dock, calling back and forth with her husband. Antonio shouts his greetings down to the captain and gives Anita a hug. The two crew are busily preparing for the late afternoon departure. They're loping back and forth between the boat and a pickup truck piled high with three day of provisions, storing the staples below deck. As the open cartons pass by, I note only necessities such as coffee, sugar, rice, crackers, potatoes, and canned fruit. Fresh fish will be the main course. Before they can push off, they must fill the fuel and water containers and organize the nets and lines that have been kept in their *bodega*—storage space—at the caleta. For Cavancha folks, the wharf is their main launching platform, their sole remaining direct access to the sea.

From their perspective, it's easy to see how the development of elite clubs and condos has boxed their caleta into a corner. Antonio and Anita soon start updating each other about the ongoing wharf case. Neither of them can fathom how the clubs could presume that the caleta could continue to exist without this wharf space. Yet, Antonio tells her how Club Yacht actually tried to push the caleta off the wharf when Iquique's recreation department recently renovated it.

Growling to Anita and me, Antonio says, "Club Yacht is only three years old, but in that short time they've already claimed that we should be kicked off the wharf for having let it run down! Club Yacht lawyers are trying to kick the caleta off the wharf by using a 100-year-old legal loophole. Never mind that today's Maritime Authority and Navy control the sea and land up to eighty meters above the high-tide mark, including all structures like wharves built out over the sea.

"Their loophole claims that our renovated wharf is on property still controlled by the municipality because—like a future planned extension—we are supposedly on a solid jetty built over an islet. But look at the water running underneath our wharf! Yes, those are laughing gulls floating around the pilings. The whole thing is a political joke."

We turn quickly to follow the roar of a couple of jet skis whizzing inshore, dodging around fishing boats on moorings. Jet skis cause accidents, but as in Miami and other coastal playgrounds, they remain unregulated toys. Antonio shifts gears, unfolding an article from the daily paper, *La Estrella*, dated 15 October 1997. It's a photo of Antonio and his fellow fisherman, Santiago, captioning a story about how the law can't control boat traffic, especially if it's part of the tourism industry. He says, "We've become like Chile's indigenous Mapuche people whose lands were slowly taken away over decades." To make his point, Antonio nods toward yet another neighbor, El Sombrero Restaurant. "During the caleta's cooperative years under socialism, that place was our seafood marketplace until Pinochet and his coup stole most of our shoreline. And now Cavancha risks losing everything with these clubs and recreation interests backed by clever lawyers."

Soria, the long-term populist Iquique mayor, promises to end the contention between the clubs and Cavancha—the wealthy boaters and the working fishermen—trying to please everyone. Antonio says, "Soria's solution will be building a lovely long extension for Club Nautico and Yacht Club, allowing us to keep our current bit of wharf space running out from shore." Antonio sums it up, "What a trial trying to survive in the middle of prime real estate."

Fishermen Become Miners as Newest Salvavida

Toxic Dust

Pablo has been EPERVA's shore plant gate guard since 1982. But Pablo, an Iquiqueño, began at nineteen as a truck driver for Chuquicamata where many Iquiqueños had migrated to work. After a decade, he decided to return to his hometown, marrying and having three children here. Pablo says, "The cold climate and toxic dust—*humotóxico*—up at the mine ultimately drove me home. Besides, I was lonely and ready to make a family. I figured ten years was the maximum time to risk toxic dust in trade for good money. We were always hearing stories about old-timers who finally retired but died shortly after from the abrupt change in climate back on the coast and from their destroyed lungs."

Perpetual Seesaw between Fishing and Mining

Labor shifts into fishing from mining in the early 1960s, after the closing of nitrate mines above Iquique. The current draw up to giant copper mines keeps the Iquique economy seesawing between mining and fishing. Chuquicamata Mine above Antofagasta did this in the 1970s. Doña Ines de Collahuasi above Iquique in the late 1990s, lured youth away from fisheries with higher income.

The increase in formally educated urban artisanal fishermen has benefited them politically in their dealings with government. Now the schooled youth from Iquique are heading up into the altiplano to the lucrative new copper mines, despite considerable health risks from working at extreme high altitude. The mining companies even refuse to accept weak or older workers. The mines provide youth with *otro futuro*—another kind of future—for those with both the education and physical strength. The mining companies are not hiring brute labor, rather those with skills such as secretaries, mechanics, machine operators, and electricians.

Industrial fisheries' shore laborers tend to lack these skills, thus don't have the chance to switch sectors and head up into the mines. Mining is overshadowing industrial fishing, prompting some fisheries labor organizations to request Iquique's regional government to give them alternative training in mining skills.

Gloria, an experienced chemist who has worked at EPERVA's shore plant since the 1991–1992 El Niño, is switching careers to go up to Collahuasi, the newest and largest of the mines. "Never mind that Collahuasi sits in high isolation, without a town, in a harsh environment!" says Gloria, who is about to alter her livelihood choice drastically. For double her salary at the fishing company, she's going to endure the 4,500-meter altitude for fifteen days at a time up in the altiplano.

The Chilean mining system operates on these blocks of fifteen days on and off, with its pros and cons for maintaining family and social life. Gloria's mother-in-law promises to help look after their two school-age kids. "It's worth it financially," says Gloria, as a married woman in her thirties. She's excited by the novelty, the starry skies, and chance to grow professionally in this hot sector. But, it's the rare woman who fits Gloria's profile and is climbing up into the world of mining.

Iquique's hungry eyes are turned toward the great copper mine of Collahuasi as a generator of jobs. Only 200 kilometers to the southeast of Iquique, Collahuasi already employs two thousand workers, half of whom are full-time. The other half are subcontracted, congruent with the fishing industry's strategy to limit exposure to labor demands. Only about thirty of the two thousand workers are women—mostly secretaries, a geologist, and Gloria, our chemist.

Pozo Almonte, a once sleepy bus stop in the pampas directly above Iquique, has swollen into a place of shacks for urban poor. But it is also the site for low-cost housing for mine workers during their half-time off. Meanwhile, the mining professionals and executives are moving into those modern homes running up the hillside at the south end of Iquique.

Dangers of Mining in the Minds of Iquique Fisheries

Canadians intend to invest over US$1.5 billion in Collahuasi to mine at a staggering altitude—the equivalent of almost 15,000 feet. To date, the Canadians have invested at least US$500 million on the project, intending this mine to be bigger and far more modern than Chuquicamata.

Tale of Hubris

In Iquique street talk, besides toxic dust, the notorious altitude has become a metaphor for human hubris and greed at the mines. Ignoring altitude demonstrates excessive pride and ignorance of nature. Legends of death circulate to show this. A popular one features either a major Collahuasi stockholder from North America, or perhaps the Canadian mine president, as the main character. He's enormously rich and fat, and refuses all advice to respect the high altitude. He assumes his time is too precious. Amplifying his risk, he gets drunk in Iquique the night before his direct ascent from sea level. Spurning medical advice to ascend in stages over a couple of days to acclimatize, he blusters his way onto the company flight straight up to Collahuasi to see his domain. Upon climbing out of the plane, he clutches at his chest and falls to the ground stricken by a lethal heart attack.

I'm having tea with one of the Catholic brothers in charge of spirituality up at the mines, and in his version of this story he adds, "Nobody could find a ready-made coffin big enough for his obese body." For locals, it's a metaphor for the danger of grasping at wealth at the expense of practical thinking. In reality, if you're willing to risk health for income as a new hire, then you must pass physical exams and take mandatory courses at an Iquique university on how to function at high altitude with scant oxygen. All mine employees must receive periodic checkups. Mining jobs suit only the young, fit, and skilled.

LETTER

Hi Bo,

I am working with free divers from the city of Iquique who are settling an isolated rural caleta. Caleta Caramucho is becoming their new permanent home base, rich with their own Management Area of 1,000 octopus caves. Sounds like the Arabian Nights! Normally, divers can collect the denizens three times a year, each time new occupants have discovered the caves. Now because of the cold waters of La Niña, octopuses have made themselves scarce. It's the opposite of when the warm waters of El Niño attract the greatest number of my cephalopod friends into shallower waters, within easy reach of divers.

I drive south out of the city, in between desert cliffs and rocky coastline, to reach my Caramucho friends. These free divers are accustomed to driving along this coast, parking down rough paths and walking out to the shore, swimming through the surf line to hunt octopus, sea urchins, and other shellfish. I find foundations and a few solid buildings already up at Caramucho. They plan to total forty houses within the year. Nicolas, the president of their free diver syndicate, has been diving here all week long. I find him sitting on a board outside his new house, repairing his net bag. Like most of the houses, his still has a dirt floor. He shows me a temporary blue tent he has erected inside the cinderblock cube. Visiting only on weekends, his wife and kids continue to live in Iquique, but everyone intends to move out to Caramucho once the place gets electricity and a few other basic amenities for urban dwellers.

For lunch, my new friend Doña Lastenia, one of the first women to live full-time in Caramucho, feeds me salty tongues of sea urchins with tomatoes, and fresh fish that her daughter just fried up for the family. Afterward, she walks me along their shoreline. Doña Lastenia explains that her two married daughters are living next door. But they refuse to help her collect and prepare the seaweed that the sea tosses up on the beaches in front of them for sale. Doña Lastenia knows it's rough work. She has been out collecting it since she was a girl.

"It's called *palo* ["stick"] because the stuff grows this way," she says, and holds up her fists to show me how her hands are covered with scars from stripping leaves off stick seaweed's branches.

Nearby her drying tangles of palo, close to shore, she points out the foundations of her family's first attempted settlement from a decade ago, including twisted and oxidized bedsprings, a seriously rusted half-moon water tank, and their good old van with open eyes where headlights used to shine. . . . To me they appear as spirits of the machine world melding into the natural one as I take pix. The van had been their link for buying seafood from the isolated caletas and selling it in the city (figure 7.3). Doña Lastenia is a pioneer spirit who works hard to earn cash from this tough environment. It's not surprising that she and her daughters have been the first caleta women to move in permanently.

I'm also finally taking digital images instead of film. The plan is to have a new computer so you'll see beyond word tales.

Later beau,
Sarah

RURAL CALETAS: THE CLIMATE OF MANAGEMENT, CLIMATE, AND SOCIAL CHANGE

Initial Intentions behind Creating the Management Area System

The dramatic loco boom and "loco war" ended in a bust in the 1980s in the Coquimbo region. Afterward, in a political climate determined to control the loco fishermen, Castilla, a Chilean marine biologist, wanted to run ecological experiments. His approach aimed to create an experimental system of Management Areas. He would base them on institutionalizing the need for numerous in-depth and repeated biological studies. Castilla, given his personal expertise and interests, did not consider socioeconomic aspects of the Management Areas. He did not delve into how to ensure economic profitability for caletas so that the Management Areas could endure.

The Navy has basically stalled on most application cases, resisting ceding Management Areas to caleta fishermen. The Navy's sub-secretary moves inordinately slowly on granting leases and permits. However, at public meet-

Figure 7.3. Doña Lastenia at Caleta Caramucho with her old van for buying shellfish from the rural caletas as far south as Caleta Chipana. Courtesy of the author.

ings the Navy never declares outright opposition to granting this marine tenure. This passive resistance contrasts sharply with the experience of salmon farmers in southern Chile, who rapidly obtained leases, given their lucrative contracts and the large business interests and budgets involved.

The government, with only one hand, is helping caletas form Management Areas to mitigate depleted stocks. Perpetual unpredictable climate events, as well as resource competition with industrial vessels and neighboring caletas, take their toll on stock levels. With the migratory safety net also cut out from under them, fishermen are left more concerned than ever about the government coming through with the tantalizing offer of marine tenure rights.

Fishermen See Marine Impermanence as Part of Livelihood

Antonio and his fellow caleta fishermen are well aware that El Niño events alter availability of marine resources. Being closest to the equator in Chile, they know that their Region I is most prone to shifts in water temperature during major El Niño events. Yet for them, nature is cyclical and will rebound as part of life. But in their hearts, the power of the industry has become an entrenched fixture.

Most urban and rural artisanal fishermen around Iquique say they experience El Niños as just one more factor of impermanence, showing off the ocean's power. Perpetual marine unpredictability is part of their livelihood. El Niño events are not seen as a bad time by them; some interpret the average El Niño as a *renovación natural*—natural rejuvenation/renovation—that comes each year for a couple of months.

They observe that life comes back better afterward, in a sort of naturally occurring marine slash-and-burn. They ignore Niño hype, knowing themselves what is going on at the moment, which is what is real and counts. The artisanal fishermen think that Chilean media, especially television, does not illustrate any good news, such as rejuvenation and tropical species moving in, because it would counter the trendy story line "El Niño catastrophe," reveal the media as hype, and perhaps gain less government sympathy that can bring regional support and funding. Fishermen also tell me they expect mass media to feed off potential tragedy and exaggeration to attract viewers and advertisers.

Each El Niño Is Different, and Fishermen Accept the Unpredictability

Direct experience of the moment resonates as the reality that matters with artisanal fishermen. They agree that they cannot predict event outcomes at any level, whether availability of species or market prices. Nor will they seriously believe the predictions of modelers and the media. They accept this uncertainty as a natural and inevitable part of their maritime livelihood.

Artisanal fishermen who are shellfish divers tend not to be blessed by Niños, unless they focus on scallop banks and aquaculture where scallops seed copiously with warmer water. Fishermen overall recall how the individuality of each event bears different consequences. The 1982–1983 El Niño del Siglo devastated their inshore environment, killing off marine life from seaweeds to seabirds, and sea lions who could no longer find food.

During the major 1997–1998 El Niño event, fishermen in the rural caletas of Caramucho and Chanavayita watched seaweeds and seabirds being wiped out. Once again, sea lion carcasses washed up on the beach to rot, and many shellfish species disappeared. Initially, this event brought a temporary glut of octopus that caused a price drop on the market. The price of octopus stayed low during the cold water event of La Niña in 1999–2000, even though the divers in Caramucho and Chanavayita watched their octopuses migrate away and become scarce.

Social Change in Caleta San Marcos

Caleta San Marcos is a compelling case of a fishing community achieving social change via its own grassroots efforts. I've driven far south of the city to reach this caleta, following the coastline to my right, dry hills to my left, ultimately turning perpendicular toward the sea on the caleta's rutted track through fields of fabulous Swiss cheese–patterned rocks of all sizes.

Nothing is paved or planted, the ground is blown bare and salty; boats are pulled up among the dozen or so shacks. Together, fishermen are heaving a boat fresh up out of the sea still loaded with shellfish. I wander over to watch. A middleman buyer sits half out of his truck, smoking and waiting. We wave. Little kids are jumping around the back of his truck. And I notice several girls huddled inside one of the beached boats concentrating on a game involving seashells.

San Marcos, like all rural caletas, began as a temporary campsite for fishermen and gradually became settled by migrant fishermen as they chose to become official residents under pressure to pick one legal home base. Their wives and children followed, creating a community atmosphere. The policy to promote Management Areas in caletas is also supposed to contribute to community stability.

Some San Marcos caleta members have decided that they need to revise their budding community's ambience to become more appropriate for family life. They are looking for ways to reduce a typical migrant camp's routine use of drugs and alcohol.

I duck inside the local restaurant/bar to start a conversation, and out of the kitchen comes Tamara in running shoes and blue track suit. Bright eyes and quick smile, she joins me over tea. Tamara turns out to be one of the earlier caleta residents from the camp days, a strong-minded single woman whose children had been long grown when she set up her establishment. Tamara says, "I take the new family arrivals under my wing, helping the women build up households. They gather here to find social support.

"I roused Tomas and a few others who were leading the caleta in coordinating fishing and marketing. Now our group has taken on the novel aspect of social reform. We're siccing the Iquique police on the unruly, even though it's a long drive from the city for the cops." The recalcitrants have been mostly single males who resist altering their lifestyle, and they have been kicked out of the caleta.

I ask Tamara how they hope to continue to keep the peace in the caleta when outsiders consider caletas as wild places. Tamara says they have the ultimate answer. Tomas's brainchild was to create access to television, and thus break the boredom that they felt was at the core of the caleta's social ills.

Tamara says, "For married and unmarried people alike, TV will be the cure!" Caleta members have rallied to establish this panacea—community television—as family-focused, safe entertainment. Leading me outside and pointing up to the top of a high hill with pride, Tamara says, "Last month, Tomas and several other men climbed that dry mountain and erected that satellite dish. And we took more caleta savings and also purchased a generator and three TV sets." Now their triumph is palpable, they feel it is safe for children to live here. The nascent community is proud of being a "decent" caleta. There is still scant light pollution to block out starry nights after the blue fluorescent glow is shut down for the evening.

Buyers Stimulate Community Growth

A good road south to Chipana, built around 1985, brought big change to the area because buyers could now make it down to the three main caletas: Chipana, San Marcos, and Chavanayita. Here, the *faluchos*—the largest model of artisanal boat with inboard engine and cabin, carrying a crew of three—from urban Cavancha also found safe harbor at night. There were no restrictions for entering the hookah diving scene in the rural caletas.

The market for octopus began booming around 1988. By 1995, octopus and shellfish stocks of locate and macha were all being heavily harvested. U.S., Japanese, and European cash-up-front buyers stimulated the growth of rural caletas. So did the profits to be had from indiscriminate harvesting of shellfish, as divers responded to the market opportunities. They were intensifying exploitation to the point of depletion.

As seasonal and quota regulations were put in place in response to the boom, black markets also arose for evading them. Several caletas—negatively characterized by their neighbors as the "Wild West"—took commonplace illicit fisheries farther, adapting loco smuggling to include drugs and contraband.

But the trend toward rural communities has given rise to local efforts at social reform, as in San Marcos. The rough image of fishermen is softening in terms of their own expectations, as more women and children arrive. Schools, family life, and social order have become important caleta issues.

RETURN TO CHILE IN 2000

The year is now 2000, and I'm again traveling south of Iquique along the dusty coastal road to the fishing enclaves. It has been two years since I last explored Iquique fisheries, but I've returned to follow up the drama where I left off. The attraction is the chance to examine the story behind Chile's budding Management Areas, and with this in mind I am heading to the constellation of rural caleta settlements, Chanavayita and Caramucho in particular. I wonder what will have transpired.

Caleta Chanavayita

Men and Their Machines

Chanavayita is the largest and one of the closest outlier caletas to Iquique, only a twenty-minute drive into the city. I hear that their fishing grounds are quite fertile, but when I'm there, I experience a stark, windy small town, exposed to sharp light, shadeless except on eerie foggy days. But Chanavayita is no longer that crude accumulation of dwellings on raised hard ground by a rocky cove. Its huge tract of shoreside land set back from the highway has been filling up with rows of curtained cinderblock homes occupied by women (figure 7.4). It emanates a sense of social well-being.

Chanavayita's expanding number of kids justifies a school, and a clinic, too, so it has become a hub for other caletas. Migrants have settled down with families, their jury-rigged gear, and precious hookah rigs resting under house eaves for filter tune-ups. Parallel streets have multiplied, curving uphill to accommodate fresh houses for all the brothers, uncles, sisters, husbands, and old family friends.

Figure 7.4. Homes in Caleta Chanavayita, the largest rural caleta. Courtesy of the author.

I've come to get a feel for the place; to see who's around to talk to. My first goal is finding the house of Maria, the elder sister of Tomas, a young diver I met buying stamps at the Iquique Post Office. We'd talked about the Coquimbo coast and how divers like him are gambling on Chanavayita's fishing grounds, having moved here from the Coquimbo region. I'd told him I'd be looking for him at his sister's in a few days' time.

Three days later, I'm driving south in the early light reflecting off the sea. Whether I can find Tomas or not, at least I'll look for Maria's and just talk to whoever isn't on the water this morning. I'm wondering what amenities or features lure divers to Chanavayita in particular, and think about their sense of well-being and their hopes for a relatively inexpensive lifestyle and unregulated livelihood.

I've noticed that people are proud of their vehicles around Iquique, and that many more households have wheels compared to other regions. I pull off the highway at the caleta sign, bouncing seaward toward a mass of houses running up a slope. Weaving up and down the dirt streets for a while, I finally spy Tomas. Wiry tall, in faded T-shirt rocking with the Rolling Stones, he's fiddling with the spark plugs on his car in front of Maria's house.

"Hi Tomas!" I call. "This place is pretty big. I've been hunting for you." He waves at me, as two of Maria's boys raise dust racing their trainer bikes in circles around us. Parking, I get out and go lean next to him peering into the engine of his green car. I make admiring noises.

Another guy comes out of the house and joins us. "I'm Jaime. Good to meet you." Jaime has long dark curls and turns out to be Tomas's cousin, who's diving off the same boat.

"What brought you here?" I ask.

"I came last year because of my brother-in-law, who runs that boat, *Lobo de Maria*," nodding down toward the cove where some twenty-five boats stretch and swing on their mooring ropes. "He wanted another diver. It's getting too much for him every day, man. He's over forty."

Tomas and Jaime bend back to the job at hand, "Hey, maybe my car was driving around Miami," teases Tomas.

"I think I saw this green machine on South Beach. How much did you pay for him here?"

"Cheap! Duty-free's great. I just can't take him anywhere else to resell." He's not alone being attracted by Iquique's duty-free port of entry for used vehicles. Tomas starts ribbing Jaime about his date tonight, and they crank up the car radio, "You can drive her to the restaurant."

First Official Management Area Comes On Line

Fisheries managers of the National Fisheries Service based in Iquique happened to be locals. With their pride of place, they wanted Chile's first Management Area to become reality in their waters.

Jorge Toro, an Iquiqueño marine biologist, spearheaded this dream. Tall and hearty, built to stay submerged in cold waters, Toro had worked for a private group for the north's artisanal fishermen, the Professional Advising and Technical Support Center of the North—Centro Profesional de Asesoría y Asistencia Técnica del Norte—called CEPAAT. In 1994, Toro moved to the national headquarters of the National Fisheries Service, and in 1996–1997 was elevated to the post of national director of artisanal fisheries. When he returned to Iquique as the Region I director of the National Fisheries Service, he had a big network and the idealism to create the first model Management Area among his rural caletas. Selecting Chanavayita, Toro banded back together with his old local colleagues and university biologists, and successfully lobbied the sub-secretary of fisheries to push through Chile's first official Management Area. They based their petition on the reams of costly, detailed biological studies that they themselves were able to carry out for Chanavayita.

Thanks to these individuals' idealism and combined institutional efforts, Chanavayita won her Management Area on 15 January 1999. This was, indeed, the first Chilean caleta to have a Management Area come on line. These Iquique managers and biologists moved more quickly and with more determination, and connections to those in power, than any other group to push a Management Area through the official stages of decree from baseline studies to final management plan.

In this way, Chanavayita's Management Area became reality before any of the other caletas could achieve theirs. The rest remained frustrated by an incredible extent of government stalling. Chanavayita could now become the national Management Area model and a Chilean model for participatory management.

Of all Iquique's rural caletas, Toro and his group of Management Area visionaries had chosen to push Chanavayita, but in the end, Chanavayita's big, shifting population of hookah divers couldn't create and sustain a unified front. Chanavayita's leaders also couldn't control the few members who were *gatos negros*—black cats, slang for thieves—raiding their own grounds and those of their neighboring caletas, also slated to gain Management Areas.

Problems in This Management Area

For some, distaste for being regulated hadn't ended with the Management Area system. I meet several of the biologists and managers who'd worked together on creating this Management Area to hear their opinions, and in hindsight they regretted choosing Chanavayita. Perhaps this Management Area was doomed from the start.

Pitufo, one of the field biologists, says, "Chanavayita's divers are lazy. We made a big mistake by handing the caleta the Management Area without any effort on their part, without a well-organized syndicate, and without the divers having to put their own money into it. The state subsidized all our studies. I think roughly only 40 percent bother to respect their own Management Area plan. Some members blatantly rob the Management Area. The syndicate doesn't even post guards because members don't volunteer their time, and they won't blow the whistle on each other anyway. Most fail to pay the monthly caleta dues (US$3 per month), though they'll contribute octopuses, and the women hold bake sales to raise cash. They just didn't earn the Management Area so they don't fight to make it work now."

When I go to talk about these problems with Toro and Vicencio, another biologist who worked in the Professional Advising and Technical Support Center of the North, both men shake their heads. Toro describes how Chanavayita was their choice because of the caleta's physical, biological, and economic attributes, but in retrospect, he rues the caleta's lack of social cohesiveness and the weak leadership among the fishermen and their families. I note that they didn't consider the participation of caleta members in the Management Area process. He blames himself because their decision making and lobbying efforts went from the Professional Center to Valparaiso and back to Iquique.

Toro calls the decision to place the first Management Area in Chanavayita an "error." He calls the caleta's members, "children," and their director "very impulsive." He adds that they are neither prepared nor clear in their goals. He says, "Caleta members lack solidarity. The Management Area has been a 'bomb' because of these kinds of conflicts that burst into the open in under half a year."

Vicencio agrees, "These social factors are behind the 'failure' of the Management Area and lie at the root of the new tensions arising between Chanavayita and Caramucho."

Caleta Caramucho

Shrine

After a good half-hour drive south of Chanavayita, I roll up to a new settlement. A bit closer to the shore a couple of rusted vehicles and a water tank are perched on the bare, cracked-cement foundations of abandoned homes by the Pacific. Along the beach, divers have posted a wooden sign for their future Management Area quite near fresh offerings of plastic flowers at the shrine of the patron saints for miners and workers. Resting inside an open-faced cement box overlooking cold ocean waters are the images of San Lorenzo of Tarapaca, Carmen of La Tirana, and Region IV's dear Niño of Sotaqui of Ovalle. This Niño—the Christ Child holding the globe in one hand—has been seen by fishermen walking in the moonlit desert.

Free Divers Desire a Management Area in Their Own Caleta

Difficult to negotiate in rough weather, Caramucho's rocky nearshore terrain makes it correspondingly idyllic for octopuses. Tucking into the rocks, they make myriad lairs. The caleta is being settled by the tiny kinship/friendship-based association of urban Iquique free divers: Sindicato de Trabajadores Independientes de Buzos a Pulmón Costero "Iquique." They've unified during the course of three generations, working together, sharing a common bond of place. This association, with their new caleta status, has been asking for a Management Area, claiming Caramucho as one of their historic free-diving grounds. Octopuses, sea urchins, clams, and loco are the four key stocks inside Caramucho's proposed Management Area undergoing the necessary lengthy studies.

In contrast to Chanavayita, as well as most other caletas occupied mainly by migrants, such as Puerto Aldea (chapter 5), Caramucho is being populated by a relatively small group that's unusually interrelated and cohesive. This gives Caramucho free divers a big advantage in terms of cooperation and cohesive decision making. These deep, long-standing ties lead them more often than not to act together and trust each other, a distinct advantage when it comes to organizing and reaching consensus for maintaining a Management Area.

Of Caramucho's fifty-odd association members, only 10 percent came from other regions. They didn't migrate to Iquique as fishermen. Rather, they arrived either as children or for compulsory military service—Iquique being a major military center for the Navy and Air Force. Only men are official association members, although their women actively participate through them. All members started out as Iquique free divers, living in Iquique. They would explore the coastline, wading out from shore at various points within the city and beyond, down to Caramucho in particular.

The association, eager to consolidate their caleta, invited the original pair of residents to join—two men who still live in a shack they built back around 1980. As Caramucho's first and longest-term occupants, these guys continue to support themselves by sifting sand for sale to construction companies. Now they have the honorific title, *socios cooperadores*—cooperative partners.

Most of the free divers still live in Iquique with their families, driving out to their caleta to harvest octopus and shellfish. By mid-2000, ten houses at Caramucho already have full-time occupants. The population is twenty-five strong. There are eight divers, six accompanied by their wives. The caleta women adamantly tell me that they're *amas de casa*—housewives. Despite their pioneer lifestyle, many cling to the middle class, urban image of the family where men are pictured as the providers. Yet this rural existence demands that women be partners in caleta work.

The mini school bus from Chanavayita fetches Caramucho's ten children. They must share classrooms and playgrounds with the offspring of their parents' giant rival caleta. Meanwhile, three teenagers stay in Iquique to attend high school. Caramucho's biggest extended family belongs to Doña Lastenia and her husband Kaupolican. They have spread out into four houses. Their life history epitomizes how caleta communities crystallize.

Life History of Doña Lastenia's Family

Doña Lastenia, as matriarch, affectionately refers to her husband as *mi patrón*—my captain—while helping me understand their life among the rural caletas raising a family. In turn, her husband, Kaupolican—the namesake of a powerful Mapuche warrior from the time of the Spanish conquests of Chile—calls her *mi vieja*—my old lady.

On a day I'm there for lunch, Doña Lastenia has spent the morning's low tide collecting seaweed. She's cutting and drying it for sale, besides managing her household and overseeing her three children's families in the houses next door.

This couple were both born and raised in Iquique and have worked as a team throughout their married life. "When we first married," says Doña Lastenia, "I remember peddling Kaupolican's baskets of limpets, clams, and sea urchins around town, shouting my wares to get customers."

In 1968, they decided to move 75 kilometers south to the far reaches of Pabellon de Pica, making their first pioneering effort in a cove that was not yet a fishing community. "Besides one other lone man, we became the sole residents, living for free on the shore. In those days, forget trucks bearing drinking water or the wealth to run generators. It was a rugged existence with our little kids. To earn a living, we started driving our van up through the caletas to buy and sell fish and shellfish. Twenty-two years later, in 1990, we moved the family in closer to Iquique, settling here at Caramucho. There were already three or four houses on the coast, none of them with land titles. So, we staked out our own place, put up a home, and continued driving our van and working as middlemen, buying and selling seafood, as far south as Caleta Chipana, the caleta furthest south of Iquique, down near Rio Loa." says Doña Lastenia.

Today she shows me the cinder-block houses popping up half built, as we walk Caramucho's coast. Wind is whipping hair in our eyes and tumbling seaweed along the shoreline. Grinning, Doña Lastenia points my gaze over to the rusted-out body of their old van, part of the historic detritus of Caramucho, confiding, "It's great living free, wrapped in waves and salt, by our own strength."

Residents of Caramucho

For years, when they were based in the city, Juan "Chino" Piérola, a craggy charming character, was president of the free divers' association. I find him at Caramucho, dressed in his signature heavy woolen navy sweater against the winds. In his early fifties, he sometimes comes out here to dive. Chino's high cheekbones are a legacy from his mother, a descendent of Iquique's numerous Chinese mine workers back in the nineteenth century. He realizes that many association members wish he'd return to the presidency to protect their interests as they develop the caleta, fighting to finalize their Management Area's pending legal status.

But Chino tells me he has become too engaged in working for the fishing industry as a net contractor to have time to run the caleta. "However," Chino says, "I'm building a second house out at the caleta." Investing in a caleta home is more than symbolic, he reminds me. "It means I'm backing our Management Area claims, and betting on the success of the government system that's trying to anchor us into single settlements."

The current association president, Nicolás Juduvish, is (not surprisingly) also building a Caramucho house. Nicolás, ruddy and fit in tight striped jersey and sweatpants, is in his prime as a diver. He's an important part-time resident, backing their bid to have a Management Area. Nicolás comes out every week, sleeping on a mattress in his fully erected two-person nylon tent, nested within four walls on a dirt floor. Like most of the thirty houses under construction, his is a slow work in progress.

"How often do you and the other divers sleep out here?" I ask.

"Most of us only sleep over, camping style, on the weekends. It's easier than commuting back to the city, and our families can join us. But we can't expect our families to move out here yet, without amenities," Nicholás says.

The weekend population balloons with divers and their families sleeping over. I see how Caramucho's spartan conditions are too similar to camping for the average female urbanite and their kids. But for Doña Lastenia, who is long accustomed to the rhythms of caleta life and prefers to be by the sea with few neighbors, this lifestyle has already been altered. Besides having a school to attend in Chanavayita, Caramucho receives a free weekly delivery of drinking water, thanks to the Iquique municipality, which trucks in a ration of 7,000 liters. Even Doña Lastenia's extended family is cultivating a more comfortable caleta lifestyle. With a subtle, proud toss of her head, the matriarch adds, "We're the only house in Caramucho with a private generator."

Aging Brings the Progression from Diving to Fishing

Kaupolican is one of Caramucho's senior free divers. Like other men his age, he is shifting from diving to fishing. "I've bought a boat instead of trusting my fins to kick me out to the octopus grounds. It's easier on old bones working from inside my boat," he says. Direct contact with Chile's bitter waters is intimidating, even with an extremely thick wetsuit.

The Chilean Navy cuts off diver permits at age forty-five, so older guys are venturing outside the law when they swim to harvest seafood. Free divers have to be tough enough to tolerate low temperatures for extended periods of time, although many keep going long after turning forty-five. Their legal option—though perhaps not economically feasible—is to switch to fishing from a boat. Already three of Caramucho's older divers have invested in boats. Two of these captains fish, but the third still wants to assert his manhood by blending fishing with diving.

Inside the house, Doña Lastenia is supervising a thick seafood soup as a couple of the younger divers and Kaupolican are discussing how older divers may shift out of the chilly waters and up into boats if they can afford to. Spooning out warm soup, he tells them, "When your bones ache like mine from cold, you'll see why buying a boat is better than quitting the life."

Beto, the smaller diver sitting next to me, nods yet pokes fun at Kaupolican for taking refuge in a boat. Then he agrees that he'd switch too, when his constitution can no longer fight the cold. They are coming to understand how this gradual trend will alter the gear composition and management of Caramucho. It will add boats to the registration list, and also change the list of fishermen who are allowed to target each commercial fish species. But they feel that their Management Area would remain the divers' domain.

Octopus Cornucopia Being Transformed into a Management Area

Around the time of the 1997–1998 El Niño, the free diver association began the process of establishing a Management Area. First, they contracted a private marine biology consulting firm from Iquique to carry out all the mandatory evaluations. These scientists hold workshops every six months with the association members, who regularly attend. To determine appropriate harvest quantities within the Management Area, the biologists maintain the trust of the divers by incorporating them into the six-month stock evaluations.

Emilio, one of the biologists, confirms Caramucho's cornucopia inside the Management Area. He describes how it's located on a seafloor, honeycombed with tiny caves and crevices. Octopus diving at Caramucho takes comparatively little effort, given that it's a densely packed, relatively shallow area. In this perfect environment for octopus, they've counted at least 1,000 occupied lairs. Octopuses love to bunch up together—called aggregating—in lairs, just as they will in traps.

"Caramucho turns out to have the perfect ecosystem for octopus, which are especially fast-growing and fecund here," says Emilio. Octopus females here are producing multiple strands of eggs that quickly hatch to repopulate the coastline. Divers see that if they wait three months after harvesting, the emptied lairs attract new denizens. During the warm water El Niño events, octopuses thrive even more. As additional gravy for the caleta, the Caramucho octopus is widely regarded by consumers and biologists alike as the tastiest and tenderest, thus commanding a top price.

Octopus Fishing

Juan emerges from the water wearing his thin worn black wetsuit and hood, the essential weight belt threaded with lead around his narrow waist. He's got his fins tucked under one arm, and is shivering slightly as he pushes his full face mask up onto his head of black hair. Spotting his pal Ricardo and me onshore, Juan beams and waves his net bag full of limp, blobby octopus bodies. I watch one suckered, pale tentacle hanging out and swaying as he approaches (figure 7.5).

Underwater at up to fifty feet, Juan has been able to spear or hook and pry each octopus from its crevice, wrestling them one by one in a suspended, upside-down weightless dance. Popping up to the surface for a breath in between, floating with his head back, face skyward, Juan takes in the brilliant blues in contrast to the water's midnight hues as he breathes and relaxes for a couple of minutes, recouping his blood's supply of oxygen.

Entering a semi-trance, human and water, he becomes a sea lion with the last three deeper breaths just before doubling over at the waist and ducking back down. One hand is holding his nose all the way to the bottom as he blows against it steadily, releasing pressure with every meter deeper into the darker world. Juan is kicking rhythmically, powerfully, and peacefully, conserving oxygen for bottom time, checking out the next lair. Familiar with the cave terrain, Juan knows the lairs' layouts. On his second descent, he peers into another and spies arms rapidly retracting and skin flashing camouflage, but too late to hide successfully. He wins. Then, tucking in his chin, he undulates smoothly in several wide-sweep kicks back to the surface light. No bubbles have escaped; no precious oxygen is lost or wasted in this hunt.

Juan, at thirty-six, is a fine free diver, feeling like a sea lion. He has practiced this art since adolescence, hitchhiking with his uncle down to Caramucho on camping/diving trips. He'd loved the desert's black night stars and cooking over a fire, having escaped urban life for a bit. During the day, they'd swim out throughout the surf to reach the field of lairs, scouting and hunting.

Figure 7.5. Free diver emerges along the desert coast with his octopus and speared fish. Courtesy of the author.

Juan, heading out today, feels that same sense of transformation flying into this realm that's dark, yet rich in color fields denied him in the desert land above: soft corals, tunicates, sponges, and others flashing oranges and yellows, reds and greens.

Occasionally a bull sea lion torpedoes by, checking him out. Female sea lions, blowing bubbles, twist and loop around him, imitating his own gyrations in prying the octopuses loose. Juan's spear strikes its target, and an inky plume pours from the hole as he works the octopus loose and pops it into the net back at his waist. This is the moment when a hungry sea lion might try to snatch a bite. Back at the surface, Juan recoups for the next descent. After hunting for an hour, up and down, he turns toward shore, elated with his success, but having expended enough precious body heat for this round.

Gaining Land and Water Rights

For several decades without land rights, a few divers like Kaupolican and his family have dwelled periodically down by Caramucho's cove. At the beginning of 2000, the Iquique free diver association claimed land rights as well as marine tenure via their decreed Management Area. By moving into Caramucho, they argue, they can guard their Management Area from outside poachers. They know that building up the caleta—shifting their residences from Iquique—is all part of the political game to secure the Management Area.

To proceed, the association asks for caleta land via the ministry dealing out land assets—Bienes Nacionales. They receive title to coastal land above the eighty-meter mark, divided into sixty-four inscribed house lots. All those who do not own homes (e.g., those who only rent or live with relatives in Iquique) qualify for one of these free housing lots, like Doña Lastenia, who has never held a title. This boon in land ownership is a major windfall and incentive to move to the caleta.

The association also asks for community construction aid, as well as state funds to build a wharf. But these supplementary requests feel far less important to the Caramucho people than gaining property rights. Besides Caramucho's ten complete houses, another thirty are in various stages of construction. This leaves room to accommodate future homes. Although Caramucho's Management Area is "decreed" as the first step, we see how anther decreed area in Puerto Aldea (chapter 5) is not allowed to harvest, despite having managed their resources to reach commercial size.

The official excuse given Caramucho for stalling their final legal status, is the die-off of a number of species during the 1999 La Niña event. Because of the delay, their costly management plan developed with the consulting biologists is no longer considered up-to-date and requires a reevaluation of all shellfish species to be harvested in the Management Area. Given perpetual marine changes, this type of demand placed on all Chilean associations seeking Management Areas feels absurd to them and disheartening, like being tantalized with a Sisyphus-like task.

Nicolás, as association leader, is particularly frustrated, saying, "If we need another set of biological reports with each new sea change, it's an excuse to keep our area in limbo forever. We've always lived with oceanic changes, but these demands to nail down a baseline for our area go against the reality of ever-changing water. It gives us the feeling of being caught up in a long-running soap opera, a never-ending management drama."

Asking him if given all this, they intend to go ahead and invest in yet another set of studies, Nicolás says, "Yes, our territory is rich. We'll guard it and fight to keep it, even if we have to pay for more damn studies."

Gatos Negros: Conflicts between Neighboring Caletas

Clash between Caramucho and Chanavayita

From the viewpoint of Toro, who originally led the selection of Chanavayita as the home of Chile's first Management Area, Caramucho would have more potential for becoming a model Management Area. Toro says, "Caramucho's Management Area is rich enough in octopus to make it worthwhile in the long term, once more members move out to the caleta. Caramucho's shared idea is to hold group harvests inside their Management Area. Unfortunately this is not even being considered by Chanavayita, where we handed everything to them for free.

"They are committed and invested in a beautiful way, respecting and guarding their own area. What a recipe for success and what a sharp contrast to some Chanavayita caleta members who try to rob their own Management Area as well as Caramucho's!" says Toro. He corroborates that Caramucho's association has already set up beach guards by their Management Area. And what tops it all, in their fight for official status, each member pays for a percentage of the expensive biological evaluations.

The fifty-hectare Management Area at Caramucho draws the envy of their giant neighbor, Chanavayita. Chanavayita divers, despite having one of the largest Management Areas in all Chile—250 hectares stretching 2.5 kilometers south of the caleta—balk at being limited to their own turf. Thus divers from the two caletas are clashing over Caramucho grounds.

After Toro and the Region I government supported Chanavayita in becoming Chile's first official Management Area, they began processing the Iquique free divers association's application for a Caramucho Management Area. This sent up furious protests from Chanavayita. How could they be kept out of nearby rich underwater octopus territory? What an outrage! They are petitioning the National Fisheries Service to rescind the Caramucho Management Area and to return the area to common use.

Already in 1995, Chanavayita had claimed "historic use" of the Caramucho area. Government mulled it over and didn't act. Even if everyone dived Caramucho's octopuses in the time before caleta settlements, the Iquique free divers were residents of the region, based in Iquique. From regional government's perspective, their citizens have priority, being here in a more stable way than divers who migrated here from other regions. Thus, Chanavayita's argument didn't fly. Instead, poaching and ill will soar.

When the struggle started, a few free-diver-association leaders went before the Chanavayita hookah divers. Limón stood up, collar open, gulped down some water and promised, "You can continue to harvest octopus in our Caramucho grounds, but only farther out in deeper waters."

They had reached an "accord of words" or gentlemen's agreement. But back at the free divers association assembly soon afterward, everyone was ranting against Limón and their other leaders. Backing down and overturning the pledge to share octopus, Limón said, "We're the true local divers with a seventy-year history of fishing up and down this coast from Loa to Arica."

This became the revised message to the Chanavayita leaders in attendance, who saw the futility of arguing, given the meeting's black mood. Besides, they fully expected Limón and these urban free divers to fail at establishing a rural population in Caramucho, and at finalizing a Management Area. Chanavayita leaders feel that in the long run, their own divers would remain, and would once again be able to harvest the octopus freely.

Black-Cat Bandits from Chanavayita Rob Caramucho Waters under Darkness

Meanwhile, Chanavayita divers like Hector, who had commonly worked the area in front of Caramucho, make a calculated shift to night diving to avoid Caramucho's new guards. Out of this Chanavayita group evolve seven die-hard robber boats manned by fourteen divers. Locally dubbed the Black Cat bandits—*Gatos Negros*—Hector is proud to be among them. Mocking Management Areas in general, the Gatos Negros steal from their own home port as well as Caramucho. Under cover of darkness, Hector and his band slip overboard, hookahs in mouth, descending with waterproof flashlights to seek out octopus. Prying them out of rocky crevices and depositing them into the shadowy boats above, the men reap quite a load in one night. By the time Caramucho guys onshore spot them, and start shouting and fanning flashing lights, they regroup on the surface, revving engines, and bolt. Hector is laughing, shivering a bit on the ride home, and counting their booty, knowing it will take a full twenty minutes before Caramucho people can drag their own boat into the water to give chase. It's an easy game.

Hector and the Gatos Negros kick into high gear during octopus season, plying the shadows most nights. In the summer of 2000—during the legal closures for octopus and sea urchins—their bandit group migrates far south to Region X to illegally dive the lucrative loco. These renegades, who cannot be controlled by their own caleta leaders, ignore all aspects of management. They do everything from robbing local Management Areas, to continuing to migrate outside their caleta's region and targeting the contraband loco.

The respite for Caramucho and Chanavayita lasted the four months the bandits were away. When they returned home more brazen than ever, their first raid took place in broad daylight, more audacious than ever. Before the Gatos Negros headed south, Caramucho leaders had filed complaints with the Navy and the port captain. The Navy actually did confiscate a few wetsuits from the divers to punish their thievery.

"Small consolation!" says Kaupolican, "Basically, policing the Caramucho Management Area falls to us. Thieves have to be caught in the act to be fined, and the agencies in Iquique responsible for enforcement are just too far away to have a prayer of catching the Gatos Negros red-handed. It's like the industrial trawlers in diver drag."

Technically, a Management Area must pay for its own guards. This is true once a Management Area is decreed yet far from being fully approved and allowed to harvest to earn income. Caramucho, therefore, bears the cost of guards without a source of cash. Members are volunteering time on rotating shifts, but the on-site population is so small that men can only do their best fanning flashlights over the sea, and screaming in agitation when they sight culprits. Perhaps they try to drag a boat into the water in time to give chase.

Mostly, they have to seek retribution in town, in a direct parallel to dealing with industrial vessels along their shore. Caramucho leaders keep meeting with the National Fisheries Service, the Navy, the port captain, and the *intendente*—regional governor—all of whom have overlapping responsibility for enforcing regulations, pointing fingers, and hoping that the other will perform the duty.

The extended families of the Gatos Negros are spread throughout Chanavayita, making it socially onerous for their caleta to expel them. In truth, Chanavayita leaders are unable to monitor, let alone control, the activities of their bad eggs. Caramucho men call the bandits bold—*atrevidos*. But, the Caramucho women up the ante, calling Chanavayita women *groseras*—vulgar—and accusing the Chanavayita Gatos Negros of hiding behind their women's skirts when officials come to investigate.

Caramucho feelings run hot and, except for the shared government school facility at Chanavayita, they have little consistent social interaction that could begin to bond the groups or lead to cooperation on the robbery issue. Chanavayita remains the giant neighbor with stores, bar/restaurants, and churches, but its people—drawn from Chile's long coast—are relative strangers, perhaps for the first generation, and unorganized in contrast to the tiny homogeneous free divers association forming Caramucho.

CONCLUSIONS

The 1991 Fisheries Law's antimigration policy, which is designed to anchor fishermen to specific caletas within one region, has set the fishermen up to disrespect and disregard the law. To date, a number of artisanal fishermen have still failed to register with a caleta. Some of them tell me this is out of feeling intimidated and disliking paperwork. Nardo is one of them. He says, "I keep fishing in hopes that, given my fishing record, the government will grandfather me in." Like Nardo, many expect this treatment down the road.

Although Region I fishermen can no longer migrate legally, caletas are full of guys who regularly return to distant home ports. But during the rotten fishing year of 1999, most returned home for a Christmas visit without trailering their boats, not because of the law, but because they decided it was not worthwhile to fish with friends and family under poor conditions. Fishermen—after their cooperative heyday in the socialist pre-Pinochet era—keep their heads down and stay rather apolitical, avoiding the system more than anything.

Only by the mid-1990s did the government finally begin consulting directly with artisanal fishermen organizations. Although the government still may not listen carefully, at least artisanal fishermen are growing more organized in their ability to communicate their needs to the government.

Local urban leaders attend the annual meeting of the national artisanal organization—CONAPACH, Confederación Nacional de Pescadores Artesanales de Chile, National Confederation of Artisanal Fishermen—although they find it difficult to elicit help from this level, which stays out of regional disputes. But Iquique's urban leaders encourage rural caletas to join together under their Region I Artisanal Federation. Leaders hope the federation movement can gain power in conflict resolution when caletas clash over Management Areas.

As they pushed Chanavayita's Management Area forward ahead of the other rural caletas with less fluid migrant populations, the band of Iquique managers and marine biologists led by Toro admit that rural caleta turf wars over Management Areas and land may intensify before they can be resolved. If Chile's Management Area system becomes regionally streamlined, then it might sooth conflicts over marine tenure rights at all levels.

EPILOGUE: OUTLOOK ON BEING ARTISANAL

Urban youth entering artisanal fishing today have a far higher level of education. Fishermen aged sixty-five and over rarely graduated primary school, averaging only four years of schooling. Their sons have often finished eight primary grades, and even four technical training years in middle school.

Young urban fishermen commonly complete two additional years at the institute level to become a *técnico pesquero*—the professional title for a technically trained fisherman. Qualified by the additional schooling, the industrial fleet recruits them as young captains. For better or worse, the formally educated generation tends to have trained in the classroom instead of having apprenticed at sea and worked their way up.

Meanwhile, the older urban caleta fishermen continue to concentrate on catching finfish. With the heart to work at sea braving storms, they constantly have been the industrial fleet's favorite recruits since its inception, whether as crew or captains. This flexible labor relationship between artisanal and industrial sectors is highlighted during major El Niño events. Urban caleta fishermen switch back to their own boats during poor industrial times as well as rich artisanal times, as well as after retiring from industrial fishing. Urban caleta fishermen have this significant option open to them during times of shifting resources, a flexibility from jobs in the industrial fleet.

Rural fishermen lack this historical labor advantage found in urban caletas. Besides urban fishermen moving back and forth into the industrial fleet, they can also take advantage of Iquique's swings between industrial fisheries and mining—the high-altitude, highly skilled endeavor, viable for younger men who have worked in the industrial sector.

Because artisanal fishermen see caleta life as integral with coping with random major onshore El Niño events, they resent the imposed inflexibility against migration along the Chilean coast. Finding economic alternatives inside and outside fisheries are matters of survival whenever there's a sea change. Rural fishermen rarely can venture far offshore and many only dive, but urban fishermen have bigger boats and can range out to sea to net more tropical species spreading south with the event's warming waters.

As we snack on fried *culengue* from a roadside stall outside Chanavayita, Toro surprises me. He thinks that despite the increasing scrutiny and guidelines being imposed on the rural fishermen, it is exactly their inherent flexibility, independent spirit, and ability to operate without government support that will probably carry them into the future. Legal or not, organized or not, conflict-ridden or not, the rural fishermen will continue harvesting as long as the sea produces a seafood that locals find pleasingly edible or buyers find lucrative for export.

LETTER: NORTH OF IQUIQUE INTO THE PERUVIAN DESERT COAST OF PARACAS

Dearie Alaka,

Sending warm thoughts while imaging you by the Great Lake with those Tibetan skull drums at the Field Museum.

At this moment, drums are beating outside this La Serena cyber café. I've just popped outside to check it out. The Chilean military are marching right by, flags stiff to the wind. The café guy reminds me it's the birthday of Bernardo O'Higgins, the ultimate father of this nation state. In the middle of fieldwork time in southern Peru, I'm sampling coastal life in between Chile and Ecuador. I'm very safe despite State Department warnings of Sendero Luminoso—Shining Path—celebrations tomorrow.

I'm feeling that the sexy *pirolín* dance music—the local Afro-Peruvian beat—blaring from the cars is mingling with dust and garbage by the roadside in an almost total disorder. At this moment the scene seems to be one lived free of constraints, like a cat's. On the bus coming to this Internet café in Pisco central, I played at seeing my Cargo Art take shape and jump out of the heaps of tires on the roofs, dirt lots piled in rubble, and a tiny, curled-up caramelo tabby kitten by her sidewalk vendor. One man told me he'd never move to our crisp, clean, green, controlled country. Nobody here imagines the existence of a slum made in the USA.

Night before last, around 7 PM, a sliver of a moon and Venus just above and to the left were almost kissing, casting a glow over the sea. The fishmeal fleet to the left floated on darker waters. Today, before dawn I took an Alaska Scallop Company truck far south into Paracas National Reserve. Pure desert hills, dunes, and cliffs, only visited by fishermen beyond the salt flats where flamingoes were wading.

It took two hours to reach Tunga's tiny three-house settlement where fishermen associations have scallop aquaculture partnerships with Alaska. Five Alaska employees, who are responsible for all the technological work, were busy sewing the sides and tops of multifloored lantern nets used for scallop grow-out. Many hope they'll need these if and when another boom to equal the one from the 1997–1998 El Niño does return next summer's Christmas time. Once again, an abundance of baby scallop seed might settle down and thickly carpet the reserve's bay.

Flocks of elegant seabirds in various forms took flight in pulses along the shore as I tried to sneak up on them, Mango dog–style. But she would have gotten far closer in her prime bird-chasing days.

Walking back toward the settlement by the surf I saw a white skull baring teeth . . . a dead sea lion, still in a mantle of fur-skin, but minus flesh and fat. His big flippers and skeleton were angled in such a way that he appeared human, even seductive. Back at the nets, I asked why the sea lion had died, wondering if it was because of the current lack of fish. No, he'd been hit by a boat, said one man shaking his head.

I've seen fishermen feeling sentimental about the death of great sea lions during El Niño events, but they'll try to kill one hanging around their nets. That's a competition. Over fried fish lunch back at Bent and Rosita's—

Alaska's owners—the maid described how she'd grown up eating sea lions and hawksbill turtles. Porpoise were common fare, too.

I'd love to return here when the wild boom town does erupt once again in Pisco. All say it's parties and processing scallops at night, then diving during the day. And robberies aren't common like now, since everyone feels fat and flush.

Again hoping all is well among your Christmas chicks. Here time is suspended, and I forget future obligations and "should-ing" which are harder to shake off at home. All this disappears in the engrossing conversations and senses of fieldwork. You know from your fieldwork.

Hoping we can rendezvous again, soon. Meanwhile, a desert flower from the rare El Niño rains. Throwing you fairy gossamer kisses and many salty and pink flamingo bird-feathered thoughts.

Much love,

Sarah

Chapter Eight

Conclusions

Coastal Ethnography Given the Eddies of Climate Change

During seventeen years of Latino Pacific ethnographic fieldwork revolving around eddies of change in political, economic, environmental, and social climates, I have been "fishing" for an understanding of coastal cultures. The tales in this book are the result. They unveil a gamut of forces operating at all political levels—including the most basic ones where real friendship expands understanding. Intermingling minds and hearts of people through the ethnographic lens, the book is an empathic examination of survival strategies as boom-and-bust fisheries cycles play out, especially around the great 1997–1998 El Niño—the last major event despite frequent predictions for another one.

In scenarios played out in multiple sites, we observe the common threads of global patterns, albeit with local twists. This ecological logic is potent. The multisited approach incorporates fieldwork at all levels, from the local level to the international level, running across ethnographies, taking nothing in a false sense of isolation. Two ethnographic studies I did during the same fieldwork period, north of Iquique in Pisagua (Meltzoff et al. 2001) and up into southern Peru in Paracas (Meltzoff et al. 2005) reveal similar patterns and provide an interesting extension.

This book provides a framework of common global fisheries patterns. The student or consultant can use the patterns described in this book as a guide for what might be encountered in any coastal research site worldwide. It offers a shortcut gleaned from my years of fieldwork in coastal cultures.

GLOBAL PATTERNS OF COASTAL COMMUNITIES

The following global patterns can be seen within each of the ethnographies and in many coastal communities worldwide, given case-by-case permutations. Recognizing these patterns is a critical first step in carrying out either coastal development and conservation projects or fisheries research and management.

Ethnographic storytelling in this book weaves descriptions of lives. Strong individuals who are often leaders of the various interest groups voice opinions based on their group's beliefs or concepts, which are learned at the knee of institutions, such as one's family, schools, political party, or religion. Concepts are often unconscious ideals or beliefs taken for granted as truth—comparable to the grammar of one's mother tongue that is automatic and unquestioned.

Rivalries and alliances shift, given the competing perspectives and interactions of different groups. We have learned, however, to look for certain commonalities. We watch diverse interest groups as allies and rivals, discovering their competing perspectives and charting their attempts to survive, stabilize fisheries, or manage new fisheries. Every case demonstrates people's knowledge and their "agency"—currently called human agency: the ability to make choices and impose them on the world. Conflicts along critical coastline habitats occur in different permutations. But by examining these interactions, the reader can gain a foundation, a set of themes to carry away and apply to the study of any coast.

El Niño Events and Ignoring Predictions

The climate variation known colloquially as El Niño pervades the book. During normal years, the Humboldt Current Large Marine Ecosystem (LME)'s colossal nutrient-rich upwelling system (www.lme.noaa.gov) flows northward from the southern tip of Chile through Peru. Its profusion of marine life supports fisheries that account for around 20 percent of the total world marine catch. El Niño events interrupt the upwelling system with intrusions of warm water associated with collapse of fisheries and famine among sea lions and seabirds. Events, however, simultaneously stimulate abundance in some marine species, such as scallops.

In Chile, there have been infatuating seesaws between the industrial anchovy and sardine fisheries, and between artisanal diving for the shellfish loco and octopus. In Galapagos, much of the coral that could have been enjoyed by tourists (e.g., inside Devil's Crown, Floreana) was wiped out by the great 1982–1983 event. Coral's recuperation is slow, but tourists come for other seascapes to see sea lions, sharks, penguins, turtles, and marine iguanas.

No two El Niño events are exactly the same, and neither are the other human-environmental interactions with climates of change. Yet patterns emerge from the ethnographies in this book that explain why fisheries sectors still do not rely on El Niño predictions by NOAA and other modelers interpreting the data. Indeed, in fisheries they are considered frequently inaccurate. Conflicting and unreliable El Niño predictions on more than ten websites offer competing, contradictory, and often faulty scientific predictions. And predictions lack the critical information fisheries require on the intensity, location on- or offshore, and duration of events.

Fishermen read El Niño predictions in the media as providing drama to sell news. They know that media, by its nature, loves stirring excitement by predicting disaster. To date, fishermen and the industry in Chile and Peru lack more serious options for knowing the future. When the science of modeling El Niño events improves, the predictions will gain credence, and their use will increase. But artisanal fishermen probably will continue trusting their in-the-moment flexible livelihood, taking what comes as long as the ocean itself remains alive.

Being in the Moment with the Inevitability of Change

Among fishermen there is an acceptance of inevitable environmental changes and surrender to surprise and the lack of control. Fishermen accept that the sea just as easily withdraws as yields wealth. Artisanal and industrial fishermen realize major El Niño events will inevitably occur, but they do not plan in advance. They see each approaching El Niño as an individual event, producing its own unforeseen set of outcomes. Fishermen believe their own eyes in the water, trusting immediate on-the-ground tacit knowledge, especially when El Niño–induced conditions and species shift before their eyes. Because El Niño events alter predator/prey scenarios, fishermen are left to harvest the newly available configurations of marine life. Their own visual confirmation in the water gives both artisanal and industrial fishermen an adequate ninety-day lead time to prepare for an event.

Fishermen conceive of the ocean as transient in her bounty, and classically have expected to have to shift grounds to find stocks as the bounty changes, and as markets boom and bust. In Iquique (chapter 6), industrial fleet managers, similar to those in aquaculture, wish to see for themselves in-the-water changes before investing in action. The industry has modified its net preparedness after being seriously burned by an El Niño event that did not duplicate the outcomes of past events as expected.

A paradigm shift in coastal management has become all the more imperative, given intensifying human-caused environmental changes—the anthropogenic factors that dominate our age.

Survival Strategies

Fishermen worldwide employ triple survival strategies: migration to better grounds, targeting different species as the ecology shifts, and turning to other types of work as a fallback. Fishermen do not starve from El Niño events, unlike farmers whose entire crop is rooted in place and can be wiped out. As much marine life shifts location to stay within their survival temperature ranges, ecology alters. Predator/prey scenarios reconfigure, shifting types of marine life for humans to harvest. Certain species, such as scallops, respond to warmer waters by producing profusions of eggs, while anchored, habitat-forming species, such as kelp, die off. Although fishermen may lose certain local, lucrative species during El Niño events until stocks recuperate, they endure by migrating to other waters, joining other fisheries, or working on land to earn a livelihood. Fishermen, even in good times, routinely maintain work options—a global pattern referred to in marine anthropology as "occupational multiplicity" (Comitas 1964).

Migration: Boom-to-Bust Adaptability in Times of Change

The migratory livelihood survival strategy has been primary, reducing vulnerability to everything from overfishing—allowing your grounds to recuperate—to times when a major El Niño event alters availability of target species. Although fishermen may lose some local, lucrative, highly marketable, species during El Niño events, the men temporarily survive by migrating to other waters, joining other fisheries, or working on land jobs.

The gold-rush mentality of migrating to take advantage of a boom is another time-honored practice. Fishermen seeking the best livelihood spread the word rapidly among themselves via cell phone, shifting grounds and even gear type and methods while the boom lasts. Some men will permanently settle and marry if opportunities seem better in the new place and love calls.

To take advantage of high market demand, buyers often initiate boom-time migration by recruiting their own harvest teams. At the promise of cash, strapped urban dwellers can also be tempted to travel far afield from familiar city life to attempt difficult and novel fishing jobs.

Fisheries as a Social Safety Net

As part of their survival strategies, fishermen will shift their target species, their methods, and their livelihood. Fisheries preserve a prelegalistic cultural concept that the lifestyle provides freedom as much as a social and economic safety net. When a fishery collapses or times are tough economically, managers may make conscious exceptions not to enforce certain regulations, as a kind of social welfare, looking the other way as long as fishermen don't fight among themselves. Fishermen and managers both hold the belief that unorganized, random urban poor (i.e., those who fish by wading that requires no gear, seen as a type of gleaning to make ends meet) can sustain themselves outside management regulations.

In the name of creating "stability" for fishermen by promising more predictable livelihoods, different interest groups have promoted tourism, aquaculture, or climate modeling. Government and NGO advisers propose tourism to wean fishermen away from fisheries and thus reduce fishing "effort." Yet the necessary capital to make the transition to high-end tourism is rarely bestowed. Fishermen with scant collateral have difficulty obtaining loans. Regular banks classically see them as high risk. Therefore, fishermen are prone to becoming false fronts for outside investors who possess the capital to take advantage of government plans aimed at reducing the number of fishermen by exchanging fishing licenses for lucrative tourist licenses.

Sidestepping Legal Regulations

Another survival strategy involves sidestepping regulations. Fishermen are prone to disregard management regulations that they see as economically hindering their livelihood in the short run. They also mistrust top-down resource management that is frequently one or two steps out of synch with their own in-the-water experience. They feel justified in ignoring or lobbying against such regulations. Ultimately, they will heed their own needs rather than regulations they deem illogical and ill-serving. Playing politics with the industry as best they can, fishermen groups, through their leaders, will switch between being allies or rivals, both sides stretching the laws and taking advantage of enforcement stretched too thin.

Fishermen will point out how biological data used to formulate regulations are out of phase with what they see in their particular underwater ecosystems. They are indignant when managers use place-specific growth rates for a species despite enormous ecosystem differences. Fishermen are also acutely aware of how policies formulated using catch-landing data do not reflect what is underwater, and thus are unfair by their standards. They are not vested in deferring to regulations based on complicated mathematical models, divorced from their own visual understanding.

Regulators tend to resist addressing this problem of top-down management. They could codesign another set of management tools with fishermen, using simpler, more visual clues. Redesigning management monitoring tools to use comprehensible math and visual monitoring with the fishermen can get far more buy-in and avert a level of sidestepping regulations. When fishermen participate in collecting visual data with biologists, their inclination shifts toward self-management.

Markets Create Fisheries

In high-priced export fisheries, marketers directly stimulate booms, and in so doing, they play into the fishermen's migration survival strategy. They can discover an underutilized stock that is precious in another's market. Then they pay people to harvest this lucrative marine product, (e.g., merluza in Patagonia to feed Spain's demand, see chapter 3). In a different scenario, marketers can jump into an area to take advantage of the temporary El Niño–induced bonanza (e.g., wild scallops thickly carpeting Paracas) (Meltzoff et al. 2005). Buyers' demands for a lucrative marine export product initiate the rush of men to harvest. The market creates the fishery.

Competing Interest Groups

Coastal real estate and tourism interests often challenge once-isolated fishing communities for their shoreline locations. Fishermen are frequently replaced by second-home owners and retirees as development causes property taxes and costs of living to escalate, as I witnessed in the Florida Keys (Meltzoff 1997). Different class aesthetics offend the newcomers who see gear or disrobing on the beach as distasteful, as in Coquimbo (chapter 4). Shores once considered out on the periphery, left to those willing to rough it, are now seen by new neighbors as littered with fishing gear and piles of seaweed in need of their class aesthetics and proper development, whether for yacht club or spa, condos or second homes. Squatters without papers can be shifted out of their coastal homes and antique structures to make room for renovation and tourism, as in Pisagua (Meltzoff et al. 2001).

Top-Down Management

Combinations of climates of change ripple through each of the chapters. Fishermen find themselves tangled in political nets of coastal policy meant to control conflicts and rivalry with real estate and tourism development, on the one hand, and with conservation interests on the other. Managers with their regulatory system aim to save marine species by orchestrating fisheries, dictating what should be extracted from the sea. But fishermen frequently resist, seeing such government interference as coming from conservationists who wish to overcontrol fishing in ways that harm their livelihood. Fishermen do not feel obliged to obey rules that run counter to their literal vision of what's in the sea and might compromise their livelihood.

Chapter 2 on Galapagos provides an ironic example: Galapagos fishery cooperatives have manipulated Ecuador's mega-tourism owners into blocking fisheries management and conservation efforts. Fishermen can easily accrue terrible international press by staging strikes or holding giant tortoises hostage. This harms bookings for the national oligarchy's yacht-based tourism. To calm the waters, they lobby friends and relatives who can rescind fisheries management. Consequently, the fishermen stop protesting and fade off the radar. Thus by blocking management, tourism interests actually help perpetuate the international conservation image of Galapagos as a haven largely devoid of humans, a place where only sea lions frolic with visitors.

I tender patterns woven around diverse narratives. The reader joins me in investigating the players who seek space in which to survive and perhaps to make innovative choices. Local fisheries leaders usually get caught up in the dialogue between agency and power, having to negotiate between their own fishermen and government managers. Their fishing groups throughout Chile are being asked by government officials to impose a degree of self-management aimed at forgoing immediate profit in return for long-term reward and marine tenure rights.

Some fishing groups can carry out self-control in harvesting, imposing stricter management than officially required, hoping to prove they deserve their pending Management Areas. However, El Niño events can throw unexpected twisters into the sea of expectations. In chapters 4 and 5, fishing groups in the bays of Coquimbo and Tongoy could self-manage their stocks. When the huge 1997–1998 El Niño event lays waste in Coquimbo Bay, yet stimulates abundance in Tongoy Bay, university biologists helping them gain Management Areas are concerned. They are afraid that the total loss of well-tended areas will frustrate those fishermen and end future self-management. Instead, fishermen, who do not expect nature to provide guarantees, accept the setback.

Honor, however, is expected among fellow humans; and the fishermen in the well-nurtured bay become angry at the government for refusing to allow them to harvest the area they have cared for. Blocking harvests until Management Areas are fully formalized, a status that appears to be forever in the future, exacerbates relations. Disheartened fishermen stop following the rules and slide back into pursuing their own logic for survival.

By exploring marine resource conflicts case by case, watching the rise of rivalries and alliances, we contemplate multivalent perspectives leading to decision making. We see managers with their biologists toying with participatory coastal management that grants marine rights, but stalling for lack of trust. Seemingly illogical stances by management induce resistance by fishermen. We see leaders of fishing groups bridling under limitations of trust and competing logic. This sends leaders hunting for chinks in a management system's armor, maneuvering behind or outside systems to negotiate a living for their fishermen. Meanwhile, governments keep generating "paper tiger" policy regulations that are toothless, ineffective, deficient in enforcement, and a temptation to ignore.

Leaders and Participatory Management

When regulations are delivered top-down, they are sidestepped if seen as unfair and uneconomical. Whenever a project is delivered top-down without being designed with the people, it isn't self-sustaining. Locals use the funds for as long as they last, then tend to let that project fold, as with the flower project for Caleta San Pedro (chapter 4).

Government and management leaders in principle have come to realize the importance of including fishermen in creating a participatory management process. Artisanal groups have shown they are willing and able to follow through to gain promised benefits, such as marine tenure rights, from participatory management.

The fishermen restructure themselves into the required associations and implement self-management to fulfill policy prerequisites. However, government and managers frequently mistrust fishermen and stay vested in and protective of elite groups along the coast (e.g., tourism, developers, and even the national Navy in Tongoy Bay [chapter 5]), leading to stalling on implementation of participatory promises. When this occurs, the fishermen in their organizations fall back into old patterns, going around regulations to earn a living.

Participatory management calls for ending top-down systems in favor of increased local control, inviting input from multilevel political agendas and authorities. The participatory management paradigm involves immediate, bottom-up grassroots empowerment. Many of us working on the coast have come to believe that if fishermen are directly involved in managing their resources, they are far more likely to see the direct benefit of cooperating with regulations. The situations studied in this book point to the desirability of participatory management and encouraging grassroots input.

Some of the decision-making power normally retained by government is shifted to local leaders and their groups for participatory management—often dubbed comanagement. Years of field experience with colleagues such as Liana Talaue McManus in the Philippines has shown that participatory management ideally involves vertical consensus building, the active and fair participation of local leaders and groups, as well as rival and managerial interest groups, formal and informal (www.unesco.org/.../liana-talaue-mcmanus-shares-her-story/).

Participatory management promotes social justice and local buy-in through the inclusion of local leaders to develop, implement, enforce, and monitor management plans. One suggestion is formulating visual monitoring techniques and biostatistical models that are grassroots-appropriate and affordable. Wolfgang Stotz, my mentor and colleague in Coquimbo, Chile, has helped pioneer Chile's system of Management Areas. He is an avid proponent of creating these user-friendly management tools so the fishermen can participate alongside biologists in monitoring their own stocks. Our detailed discussion of Chile's model Management Area system is laid out in Meltzoff and others (2002).

In Paracas (Meltzoff et al. 2005), I found Peruvian officials and biologists looking seriously to Chile's Management Area system for guidance. Yet Chile, Peru, and Ecuador still fall short of developing innovative user-friendly, monitoring techniques with the fishermen.

This book illustrates the lingering reluctance of governments to entrust fishermen with modicums of resource control, whether over artisanal Management Areas for wild harvest and aquaculture, as in Chile and Peru, or over the issuing of tourist licenses in exchange for fishing licenses, as in Galapagos. Loath to relinquish control to the grass roots, managers make excuses. They point to years of fishermen's defiance of regulations. They set requirements to repeatedly produce costly biological reports as climate conditions change, which fishermen see as a Sisyphean task. Concurrently, they may be quietly receiving messages from elite groups such as tourist developers, industrial tourism owners, and industrial fisheries.

Fisheries associations continue to be held at arm's length by their governments. They either remain excessively dependent upon outside fisheries professionals and advisers who must be hired at great expense to keep up the stream of reports, or upon the kindness and idealism of volunteer university biologists doing this research.

Participatory Management Leaders

My previous fieldwork, in Spain, the Solomon Islands, and elsewhere, led me to focus increasingly on the role of leaders among fishermen. Inspired by prospects of wholehearted participatory management, it's apparent that transformational management styles require extraordinary individuals. These leaders are not simply manufactured through outside training workshops and programs. For success, leaders need to be trusted locally. Being homegrown becomes an advantage. Leaders attempting transformational management styles are most effective when they have both local knowledge and formal education, bridging concepts of cause and effect, of how ecosystems and politics function.

A charismatic leader is more likely to have success controlling a smaller group with long-term membership rather than a larger one composed of recent migrants. Ironically, leaders of fishermen's groups—whose job is bridging relations and negotiating with those in power—endure a level of mistrust and dearth of cooperation from the very fishermen they represent.

The Future of Fisheries Management

To date, no management framework in Chile, Peru, or Galapagos has been able to generate the feedback across levels necessary to sustain trust and local compliance. Achieving consensus and a local conservation ethic that inspires compliance will entail grassroots management. Otherwise, fishing people could remain essentially outside the law, with fishing regulations unenforceable despite detailed rules on paper, lopsided in terms of their perceived logic. To develop grassroots strength, a management concept should stress the value of social capital and social well-being, including quality leadership. Decentralized power structures are crucial to coping with globalization and climate variation.

My fieldwork demonstrates that the more viable fishing communities—those that can perpetuate economic output and a standard of living in times of change—are generally characterized by focused community-development efforts and strong social affiliations. From another angle, fishing communities that have existed for several generations are more easily managed by their homegrown leaders, as compared to the newer, more fluid, fishing communities where even the leaders migrated in from the outside. We see this in sharp focus among the artisanal fishermen around Coquimbo and Iquique (chapters 4 and 7).

Governments and their managers could hire fishermen and entrust them as guards in off-seasons, like Galapagos fishermen ask of the Park in vain (chapter 2). Governments could come through on promises of marine tenure. Instead, they fuel mutual mistrust by stalling for years. Government obstruction of their own Management Area systems and special leases are examined in detail in Coquimbo, Tongoy, and rural Iquique (chapters 4, 5, and 7). Fisheries can crash economically when regulatory efforts are blatantly manipulated and gutted by powerful lobbyists until harvests become too poor to attract buyers, as with sea cucumbers in Galapagos (chapter 2). At various political levels, government and managers are characteristically suspicious of fishermen and fear a transfer of control to coastal communities, where more powerful elite tourism and conservation groups want ascendancy. Every chapter shows this.

Overall, empowering the coastal people and their leaders is a priority in creating a true multilevel partnership, hence generating a sense of shared investment in the management process. Their perceived and experienced risks and livelihood security profoundly influence their degree of resistance or compliance with management. One hopes that government and fishing people will work together to reach the point where both participate in local management, monitoring, and guarding fishing grounds out of season. This would require both sides to develop (1) livelihood models that are socially expedient and include well-being; and (2) assessment tools that reward self-reliant local stewardship of the resources. This book is a call for such partnerships; for developing tools that honor tacit knowledge, visual analyses, simple arithmetic, and qualitative data. Local buy-in on the ground evolves from facilitating this process of ongoing, design-built coordinated interaction.

GRASSROOTS EFFORTS WITHIN THE COMMUNITY

> *One touch of nature makes the whole world kin.* —William Shakespeare, Ulysses, in *Troilus and Cressida*, act 3, sc. 3, l. 175–80

I've been listening to Isabeleños' marginalized voices from the years of their dive deaths and their strikes up to their current hope for local tourism. I felt a palpable urgency to act beyond the role of ethnographer, to assist Isabeleños in their quest for grassroots geotourism and conservation education in the iconic Galapagos environment. The urgency of fast-paced change led to co-founding Isabela Oceanographic Institute (IOI) (www.ioi-galapagos.org/).

The National Geographic defines geotourism as

> tourism that sustains or enhances the geographical character of a place—its environment, culture, aesthetics, heritage, and the well-being of its residents.
>
> geotourism incorporates the concept of sustainable tourism—that destinations should remain unspoiled for future generations—while allowing for ways to protect a place's character. Geotourism also takes a principle from its ecotourism cousin—that tourism revenue should promote conservation—and extends it to culture and history as well, that is, all distinctive assets of a place. (National Geographic 2012)

When Johann Besserer took my 2005 Galapagos field course as a master's student in marine affairs and policy at Rosenstiel School of Marine and Atmospheric Science, he wanted to join me in making a grassroots institute in Isabela. Johann and I started to cocreate IOI, a Florida nonprofit. Johann stayed in the archipelago, flying back and forth to Quito and Guayaquil to work with the imperative lawyers. After five years of alliance building, legal work, political meetings, and trial courses, our political ducks were aligned.

When University of Miami's provost, Tom LeBlanc, came on one of my field courses, he and President Dona Shalala quickly understood the appeal of creating a UGalapagos program run by IOI. Our first full semester UGalapagos program began in spring 2010. Running UGalapagos programs provides IOI with financial stability for our community projects.

By keeping overhead low, IOI is able to funnel most of the income back into the community. Besides funding modest infrastructure, IOI helps the municipality run workshops, programs, and conservation events. Above all, IOI is community oriented. We depend on locally owned stores and restaurants for food, and use only island vehicles and boats for transportation.

IOI has engaged a respected local director, Sara Luz Ruiz Escobar, local leadership being a critical component for all grassroots efforts. Sara Luz knows Isabela's families through a decade of teaching Villamil's younger generation. At the Galapagos provincial level, she is part of the small pool of friends who alternate roles inside the various Galapagos conservation organizations.

Sara Luz can navigate the correct environmental course of action for IOI, down to the type of visas for our students and professors, and the fairest way to select which community projects to fund. With mandatory yet ever-changing government permits Sara Luz ensures that IOI has the necessary Ecuadorean links and follows all legalities.

Each UGalapagos student now lives in a separate homestay. Homestay families in a tiny community where everyone knows each other are partially selected for being good community role models for a conservation ethic. The municipal selection process acts as an economic and social incentive among families who wish to host students.

Immersion is successful, students and their new families bonding as they share ideas and language. They exchange views of nature and good conservation, what constitutes family well-being, and views on economic options. A number of students have already returned for visits to their host families, and the families stay in close contact with their new children/siblings.

As another component of the community experience, UGalapagos students volunteer in a project for their civic engagement course. The only environmental after-school center is happy to have help teaching tide pooling marine biology, art, and music. The municipality soccer and basketball summer camp appreciates coaching. The Galapagos National Park, too, presents projects, such as repairing tourist-site boardwalks and beach cleanups with children.

From the beginning, IOI has been providing native-speaker English teachers for the schools. Many people want training in basic English to be able to communicate directly with tourists. We are putting together a public library from donations. IOI is expanding into supporting other types of courses to help the community compete with the outside tourist interests, to be able to retain their own tourism, and to become a model for global geotourism. People have asked IOI for computer Internet workshops to tie into the Galapagos tourism network. Since 2009 with the building of a tower next door to IOI, we have round-the-clock Internet and aim to expand educational and community outreach for local businesses. IOI is planning a course in small business administration, including the use of the Internet.

Given our record in Isabela, we now are being recognized and invited by the major foundations in Galapagos to join in conservation efforts. IOI hopes to have a lasting relationship, using an educational grassroots approach to maximize community benefits, helping establish a stable social economy, one that protects both the sense of well-being as well as the archipelago's spectacular life for all humanity.

Appendix

Marine Life Names in Latin

Albacora (Chile), pez espada (Chile, Peru), swordfish (*Xiphias gladius*)
Anchoveta, Peruvian anchovy (*Engraulis ringens*)
Atún, atún de aleta azul, tuna (*Thunnus thynnus*)
Bonito, Eastern Pacific bonito (*Sarda chiliensis*)
Cabinza/cavinza, cabinza grunt (*Isacia conceptionis*)
Cabrilla, Peruvian rock sea bass (*Paralabrax humeralis*)
Camarón, langostino, shrimp, prawns (multiple species)
Cangrejo nadador (Peru), pelagic swim crab (*Euphylax dovii*)
Caracol rubio (Chile), snail (*Xanthochorus cassidiformis*)
Chicorea de mar (Chile), yuyo (Peru), nori (Japan), a seaweed (*Chondracanthus chamissoi*)
Chocha (Chile), a limpet (*Calyptraea trochiformes*)
Cholga (Chile), choro (Peru), cholga mussel (*Aulacomya ater*)
Cochayuyo (Chile), bull kelp (*Durvillaea antarctica*)
Cojinoba, bluefish, palm ruff (*Seriolella violacea*)
Concha de abanico (Peru), ostión del norte (Chile), scallop (*Argopecten purpuratus*)
Congrio, cusk-eel (*Genypterus spp.*)
Corvina, sea bass, corvina drum (*Sciaena gilberti*)
Culengue (Chile), almeja (Peru), clam (*Garisolida*)
Erizo (Chile), erizo rojo (Peru), sea urchin (*Loxechinus albus*)
Estrella de mar, starfish (covering multiple species of echinoderms)
Huiro (Chile), sargazo (Chile, Peru), alga parda (Peru), giant bladder kelp (*Macrocystis pyrifera*)
Jaiba (Chile), crab (*Cancer spp.*)
Jurel (Chile), horse mackerel (*Trachurus murphyi*)
Jurel (Peru), Pacific jack mackerel (*Trachurus symmetricus*)
Lapa, keyhole limpet (*Fissurella spp.*)
Lenguado, fine flounder (*Paralichthys adspersus*)
Lobo de mar (Chile, Peru, Ecuador), lobo marino (Chile, Peru, Ecuador), South American sea lion, lobo Chusco (Peru) (*Otaria byronia*) and lobo fino (Peru) (*Otaria flavecens*) (only Peruvian fishermen make a distinction between these two species); Galápagos sea lion (*Zalophus californianus wollebaeki*)
Locate (Chile), caracol (común) (Peru), top shell (*Thais chocolata*)
Loco (Chile), chanque, pata de burro (Peru), Chilean abalone (*Choncholepas choncholepas*)
Luche (Chile), alga roja, cochayuyo (Peru), red seaweed (*Porphyra columbina*)
Macha, surf clam (*Mesodesma donacium*)
Merluza austral (Chile), southern hake (*Merluccius australis*).

Merluza (Chile, Peru), hake (*Merluccius gayi*)
Mojarilla (Peru), minor stardrum (*Stellifer minor*)
Navaja (Peru), razor clam (*Ensis macha)*
Nori (Japan), edible seaweed pressed into laver sheets, chicorea (Chile) (*Porphyra spp.*)
Ostióndel sur (Chile), scallop (*Chlamys patagonica*)
Ostra (Chile), Chilean oyster (*Ostrea chilensis*) native and Pacific oyster (*Crassostrea gigas*) introduced
Pejerrey, silverside (*Odontesthes regia*)
Pelillo (Chile), yuyo (Peru), seaweed (*Gracilaria chilensis*)
Pepino, sea cucumber (*Isostichopus fuscus)*
Perico (Peru), dorado (Chile and Peru), mahimahi, dolphinfish (*Coryphaena hippurus)*
Picoroco (Chile), pico de loro (Peru), barnacle (*Megabalanus psittacus*)
Pinguino (Galapagos), Galapagos penguin (*Spheniscus mendiculus)*
Pinguino Magallánico, Magellanic penguin (*Spheniscus magellanicus*)
Pintadilla, Peruvian morwong (*Cheilodactylus variegatus*)
Piure, ascidian, sea squirt (*Pyura chilensis*)
Pulpo, octopus (*Octopus spp.*)
Sardina Española (Chile), sardina (Peru, Chile), sardine, South American pilchard (*Sardinops sagax*)
Salmón, Atlantic salmon (*Salmo salar*), Pacific salmon (*Oncorhynchus kisutch*), king salmon (*Oncorhynchus tschawyscha*)
Taca (Chile, Peru), almeja (Peru), taca clam (*Protothaca thaca*)

Glossary

agency (human): The ability of humans to make choices and impose them on the world, in contrast to natural forces.

alcalde de mar: *Caleta*-elected "mayor" who organizes daily fishing permits for each trip and helps enforce fishing regulations; he is the political leader of the *caleta.*

altiplano: A high plateau in the Andean region.

armador/es: Vessel owner who directs operations.

baitfish: Small schooling species used to produce fishmeal and fish oil (e.g., anchovies and sardines).

balsa: A centralized floating fishing base, which is the heart of the *merluza austral* operation. Balsa is a common soft wood (*Ochromapyramidale*), very light in weight, also used in the construction of boats or rafts in pre-Columbian times in South America.

benthic fish: Bottom-dwelling fish.

Bienes Nacionales: Chilean ministry dealing out land assets

botes: Open artisanal boats.

buzos a pulmón: Free divers. Divers who rely on breath-hold diving instead of scuba tanks or hookah hoses connected to air compressors in boats that follow overhead.

caleta: Spanish for "cove." In Chile and Peru, *caleta* is the common term for an artisanal fishing port, or fishing coves that have evolved into communities.

cardumen: Great school of fish.

capataz: Foreman. Chapter 6 talks about net repairers or weavers who are guided in technique and procedure by their professional leader—the *capataz*. In Iquique, the industrial fishing companies each kept a chief of nets—*jefe de redes*—who made the final net design decisions. Then, alterations in net design could be suggested by net leaders as well as by the captain of a fishing vessel. Today's net leader is responsible for both. He knows how to repair the massive nets by sewing new sections into worn-out or ripped sections; and how to redesign a net's size and shape, which is generally done by adding zigzags in appropriate places to deepen the net.

chamaya: Nickname for cannery workers, who earned the name of this smelly little fish by going home impregnated with its stench.

chinguillo: A long, narrow mesh bag carried to collect shellfish or fish.

El Niño (El Niño Southern Oscillation, E NSO): A phenomenon in the equatorial Pacific Ocean. See NOAA (2009b and 2009c) for a detailed scientific understanding of how this event is a warming of the ocean surface off the western coast of South America. It tends to occur every four to twelve years when upwelling of cold, nutrient-rich water does not occur. It triggers die-offs of plankton and fish, and thus their predators, like sea lions and seabirds.

It affects Pacific jet stream winds, altering storm tracks and creating unusual weather patterns in various parts of the world.

El Niño del Siglo: "The El Niño of the Century." Refers to the mega-ENSO event of 1982–1983 due to its intensity and world impact.

falucho: The largest artisanal boat, with inboard engine and cabin, carrying a crew of three.

fibra: Open fishing skiff (Ecuador).

fishmeal: *Harina de pescado*. Used for animal feed.

fish oil: *Aceite de pescado*. Often used as a food supplement.

Gatos Negros: Black Cats, Slang for thieves at sea as well as on land.

geoglyph: A drawing often made on hillsides, created by arranging earth, stones, or gravel, or by taking away patinated clasts to expose the ground.

gill-netting: *Red de enmalle*. A fishing net set vertically in the water so that fish swimming into it are entangled by the gills.

gobernación marítima: Marine government. A section of the Chilean Navy, which can be compared to the U.S. Coast Guard. It issues the appropriate fines or suspends work permits for fisheries.

hookah divers: Divers who descend with long hoses connected to a surface air compressor in a boat overhead.

Intendente Regional: Regional governor. In Chile it refers to a regional local government official.

La Niña: A coldwater event. A cooling of the ocean surface off the western coast of South America, occurring periodically every four to twelve years and affecting Pacific and other weather patterns. See NOAA (2009b and 2009c) for a detailed scientific understanding.

lanceador: Artisanal fisherman who balances on the extended prow of the boat, hurling a lance at the billfish breaking the surface.

***Ley de Pesca* (1991 Fisheries Law)**: A major piece of Chilean legislation that attempts to force migrant fishermen to settle within one region. This includes setting up registries for each commercial species.

longlining: A fishing method that sets a main line to which are attached branch lines, each fitted with one or more baited hooks or artificial lures. It can be either surface set (pelagic) or bottom set (demersal) line fishing. In Peru this method is called *espinel*.

macheros: *Macha* (clam) harvesters.

Management Area: A Chilean fisheries system area designated for a fishing association to a have exclusive use and management.

marine tenure: An ownership of marine areas, parallel to land tenure.

mariscos: Term used to refer to all types of shellfish.

nori: A seaweed that the Japanese press into sheets to wrap around raw fish and rice to make sushi.

orilleros: Those who wade to harvest *macha*. The term is derived from the word in Spanish for shore—*orilla*.

pampa: Quechua word meaning "plain." It usually refers to the Argentinean fertile lowlands; however, in Chile and Peru it refers to the coastal desert flatlands.

panga: Little open boat (Ecuador), often towed behind the mother boat for fishing.

participatory management: In fisheries management it stresses inclusion of fishermen to develop, implement, enforce, and monitor management plans and regulations. It calls for the most immediate, bottom-up grassroots empowerment, and is a step beyond integrated coastal management (ICM), where the formal government levels remain in power.

patrón de pesca: Fishing-vessel captain.

pelagic fish: Deep-sea fish.

petroglyph: A painting or carving on a rock, often on the side of a large boulder or inside a cave.

pincho: Fishing "art" or technique used around Puerto Montt, Chile, where the fisherman makes a jerking action on a line with up to five hooks to lure fish.

purse seining: *Cerco de fondo* (Chile), *boliche* (Peru). A fishing seine that is drawn into the shape of a bag to enclose the catch. This harvesting method encircles a mass of pelagic fish within a large wall of net. The large net can encompass a great school of fish. Once a school is encircled, the bottom of the net is closed, or pursed, shut. Then it's hauled in. Artisanal boats haul far smaller nets, usually closer to shore at shallower depths than their industrial competition.

recuperación de sólidos: Term used in the industrial fishing plants that refers to the recovery of solid wastes.

romería: A pilgrimage.

ruco: Chilean word for a rustic dwelling of patched-together scraps of canvas, plastic, wood, and tin.

salar: Salt flat or salt pans. A dry lake bed consisting of fine-grained sediments, primarily salt.

salvavida: A lifesaver, used literally or figuratively.

seasons south of the equator: Being south of the Equator, coastal Chile and Peru have seasons reversed from the United States and Europe in the north: September–October is spring; December–January is summer; April–May is fall; and July–August is winter.

seasons in Galapagos: Due to biannual swings in ocean currents of colder or warmer waters, Galapagos archipelago experiences just two seasons: a colder wet season that peaks around August, and a warmer dry season that peaks around January.

Sendero Luminoso: Shining Path. A Peruvian rebel group, founded with socialist ideals that has operated through acts of violence within the country since the 1980s.

Special Law of Galapagos: 1998 legislation addressing population growth by calling for strict residency requirements in attempts to end rampant illegal migration to Galapagos.

sub-secretary of fisheries: Chile's highest governmental fisheries officer.

tele: Local slang for the crew member who stays topside running the boat, monitoring the divers and their safe air supply. He follows each diver by watching their air bubbles, checks that the hookah compressor is producing the life-sustaining air supply, and helps to keep each long hookah hose running free and untangled.

varazón: Uncommon phenomenon of currents that heaves bottom life onto the beach, including scallops in Tongoy, Chile. In Paracas, Peru, it refers to massive die-offs of marine life caused by red tides or, more commonly, by untreated sewage discharge from the fishmeal plants.

veda: Closed fishing season.

References

BBC News
2010 Galapagos Islands Taken Off UNESCO Danger List. www.bbc.co.uk/news/world-latin-america-10808720.

Biersack, A., and J. Greenberg
2006 Reimagining Political Ecology. Durham, NC: Duke University Press.

Brown, K. McCarthy
2001 Mama Lola: A Vodou Priestess in Brooklyn. Berkeley: University of California Press.

Comitas, L.
1964 Occupational Multiplicity in Rural Jamaica. *In* Proceedings of the American Ethnological Society, 1963, pp. 41–50. Reprinted in Comitas, Lambros, and David Lowenthal (eds.), Work and Family Life: West Indian Perspectives. Pp. 157–173. Anchor Press/Doubleday, 1973.

The Economist
2010 UNESCO's world heritage sites: A danger list in danger. 26 August.

Galapagos National Park
2003 www.galapagospark.org/documentos/DPNG-FCD_pesqueria_2003.pdf

Gravez V., C. Segura, and A. Gelin
2008 Análisis de la situación socio-económicadel sector pesquero artisanal de la isla Isabela, Galápagos. Ecuador: Cooperativa de Pesca Artesanal Horizontes de Isabela, Galápagos and Fundación Futuro Latinoamericano.

Gudeman, Stephen, and Alberto Rivera
1995 From Car to House (*Del coche a la casa*). American Anthropologist 97(2):242–250.

Holmes, D. R., and G. E. Marcus
2008 Cultures of Expertise and the Management of Globalization: Toward the Re-Functioning of Ethnography. *In* Global Assemblages: Technology, Politics, and Ethics as Anthropological Problems. A. Ong and S. J. Collier, eds. Oxford: Blackwell.

Instituto Nacional Galápagos (INGALA)
1998 Ley especial para la provincia de Galápagos. Ley No. 6, RO/ 278 of 18 March 1998. www.ambiente.gov.ec/paginas_espanol/3normativa/docs/GALAPAGOS2.PDF, accessed 29 September 2009.

Isabela Oceanographic Institute (IOI)
2009 The Isabela Oceanographic Institute. www.ioi-galapagos.org, accessed 29 September 2009.

Jo-Frazier, L.
2007 Salt in the Sand: Memory, Violence, and the Nation-State in Chile, 1890–1997. Chapel Hill, NC: Duke University Press.

Klein, N.
2007 The Shock Doctrine: The Rise of Disaster Capitalism. New York: Henry Holt and Company.

Kornfield, J.
2008 The Wise Heart: A Guide to the Universal Teachings of Buddhist Psychology. New York: Random House.

Marcus, G.
1995 Ethnography in/of the World System: The Emergence of Multi-Sited Ethnography. Annual Review of Anthropology 24:95–117.

Meltzoff, S.
1995 Marisquadoras of the Shellfish Revolution: The Rise of Women in Co-management on Illa de Arousa, Galicia. Journal of Political Ecology 2:22–38.
1997 Beneath the Veneer of Paradise: The Struggle over Cuban Ethnic Identity and Place in Key West. Ecumene 4(2):139–160.
2001 Development and Fisheries Management among the Political Ghosts of Pisagua, Chile. Journal of Political Ecology 8:45–80.

Meltzoff, S., and E. LiPuma
1986a The Troubled Seas of Spanish Fishermen: Marine Policy and the Economy of Change. American Ethnologist 13(4):681–699.
1986b Hunting for Tuna and Cash in the Solomons: A Rebirth of Artisanal Fishing in Malaita. Human Organization 45(1):53–62.
1986c The Social and Political Economy of Coastal Zone Management: Shrimp Mariculture in Ecuador. Coastal Zone Management Journal 14(4):349–380.

Meltzoff, S., and J. Schull
1999 Miskito Ethnic Struggle over Land and Lobster: Conserving Culture and Resources on Corn Island. Journal of Culture & Agriculture 21(3):10–18.

Meltzoff, S., M. Lemons, and Y. Lichtensztajn
2001 Voices of a Natural Prison: Tourism Development and Fisheries Management among the Political Ghosts of Pisagua, Chile. Journal of Political Ecology 8:45–80.

Meltzoff, S., W. Stotz, and Y. Lichtensztajn
2002 Competing Visions for Marine Tenure and Co-management: Genesis of a Marine Management Area System in Chile. Coastal Management 30:85–99.

Meltzoff, S., M. Lemons, L. Asfour, G. Guardia-Montoya, and R. Gonzales
2005 Sustaining El Niño-Induced Scallop Booms with Aquaculture: Livelihood Transitions of Artisanal Fishing People in the Paracas National Reserve, Peru. Journal of Culture and Agriculture 27(1):1–15.

National Geographic
2012 Center for Sustainable Destinations. Electronic Document. http://travel.nationalgeographic.com/travel/sustainable/about_geotourism.html , accessed March 17, 2012.

Neumann, R.
2005 Making Political Ecology. London: Hodder Arnold.

Nichols, H.
2006 Lonesome George: The Life and Loves of a Conservation Icon. London: Palgrave Macmillan.

NOAA
2009a El Niño Theme Page. www.pmel.noaa.gov/tao/elnino/nino-home.html, accessed 29 May 2009.
2009b What Is an El Niño? www.pmel.noaa.gov/tao/elnino/el-nino-story.html, accessed 29 May 2009.
2009c North American Countries Reach Consensus on El Niño Definition. www.noaanews.noaa.gov/stories2005/s2394.htm, accessed 29 June 2009.

Pálsson, Gisli
1994 Enskillment at Sea. Man 29(4):901–927.

Pálsson, Gisli, Bronislaw Szerszynski, Sverker Sörlin, John Marks, Bernard Avril, Carole Crumley, Heide Hackmann, Poul Holm, John Ingram, Alan Kirman, Mercedes Pardo Buendía, Rifka Weehuizen

2012 Reconceptualizing the "Anthropos" in the Anthropocene: Integrating the Social Sciences and Humanities in Global Environmental Change Research. Submitted to Environmental Science and Policy.

UNESCO World Heritage Sites. www.whc.unesco.org, accessed June 2007.

Index

About the Author

Sarah Keene Meltzoff is associate professor at Rosenstiel School of Marine and Atmospheric Science, University of Miami, where for eighteen years she chaired and built Marine Affairs and Policy. Meltzoff's métier has always been ethnography. Her global fieldwork in coastal cultures spans four decades, from the Solomon Islands and Spain to Thailand and the Florida Keys. Research in the Latino Pacific for this book took place over seventeen years. Meltzoff is also chair and cofounder of Isabela Oceanographic Institute, Galapagos, a grassroots educational nonprofit for conservation.

www.ingramcontent.com/pod-product-compliance
Lightning Source LLC
LaVergne TN
LVHW101445200926
848483LV00001B/54
* 9 7 8 0 7 5 9 1 2 2 3 5 2 *